NO TURNING BACK

NO
TURNING
BACK

DISMANTLING THE FANTASIES
OF ENVIRONMENTAL THINKING

Wallace Kaufman

toExcel

San Jose New York Lincoln Shanghai

No Turning Back

Dismantling the Fantasies of Environmental Thinking

Published by toExcel Press,
an imprint of iUniverse.com, Inc.

For information address:
iUniverse.com, Inc.
620 North 48th Street
Suite 201
Lincoln, NE 68504-3467
www.iuniverse.com

ISBN: 0-595-00099-1

For these three who have given the greatest services of friendship: honest criticism, unquestioning help, and loyalty in all weather.

My daughter, Sylvan Ramsey Kaufman.

My friends Ed Holmes and Mary Jane Boren Meeker.

CONTENTS

ACKNOWLEDGMENTS

I must thank my daughter, Sylvan, who has disagreed with many of this book's conclusions, for her fair and thorough help digging for facts and documentation. We may each change our minds on parts of this book, but I trust we will not abandon our belief that facts are the foundation for opinion.

Rebecca Clay deserves credit for help at critical moments, for many editorial suggestions, and for encouragement. Annabelle Andrade, a fine teacher at Roosevelt University, made many valuable observations as she sifted through parts of the manuscript. She often steered me to ideas about religion and culture that I had not explored. I have no greater debt to anyone than to Edith Calhoun for her determined and resourceful work on many final details.

Many others contributed by their scholarship and insightful writing. I have tried to cite all the sources I consulted directly as I wrote. I am particularly indebted to a few friends in the environmental movement who have been willing to suffer my criticism and argue without abandoning our friendship. And I want to acknowledge the strong debt I owe to historians of the environmental movement.

NO TURNING BACK

CHAPTER 1

Confessions of an Environmentalist

There would be some truth in calling this a "kiss and tell" book, since it is by an environmentalist about the environmental movement. I have lobbied for the Wilderness Society. I presided over statewide environmental groups as we sued the U.S. Corps of Engineers and intervened in the licensing process of nuclear power plants. I continue to prefer wilderness to cities, solar power to nuclear power. I built my own passive solar home. I dry my clothes on the line, garden organically, and gather edible wild plants. But I do not agree with Thoreau or the Sierra Club that "in Wildness is the preservation of the world." The opposite is true. In civilization is the salvation of wilderness and of nature in general.

Over the past twenty years, I have lived in the woods, farther from neighbor or passersby than Thoreau was at Walden Pond. I have listened to nature, and because I care about it, I have also listened to the voices of the environmental movement. During the past ten years, as the book and video reviewer for a national environmental magazine, *American Forests*, I have read everything from books on ecotherapy to government policy papers. The environmental magazines I have received consumed a small grove of trees. The nature I live in and write about is quite different from the nature discussed in most environmental writing and at conferences. As a science writer, I also often find that nature for environmentalists is quite different from nature as scientists know it. These differences are one subject of this book.

The other experiences that moved me to write this book were my frequent travels in Latin America and, during the past five years, in the former Soviet empire. As this book goes to press, I will be living in Almaty, the capital of Kazakhstan. Almaty, once Alma Ata, means "place of the apple." The earliest known wild apple trees grew in the hills around the city. They still grow here. This is a rare place in our world, and that is part of the reason I am here. The other reason is that for five years I have expected to find in the ruins of the former Soviet Union reasons why human beings often behave so badly toward each other and toward nature.

Everyone knows how horribly the Communists fouled their environment. I have traveled from one end of the former Soviet Union to the other, from Poland to the Pacific coast of Siberia, from the Arctic Ocean to Kazakhstan. I am now living not far south of Semipalatinsk, where the Soviets tested more than 466 nuclear bombs, many of them above ground and within sight of towns and villages. You can blame what I have seen on communism, but it was not just communism. It was the near fatal error of distrusting individual freedoms and believing that government is the best planner and decision maker. The more I have traveled in this world of ruined landscapes and ruined people, the more I realize that American environmentalists harbor the same fear of human nature and the same dependence on government command.

Life here in the woods and travel in the world's wastelands have made a kind of intellectual sandwich for thinking about environmentalism. I find myself firmly convinced that I no longer live in the same world as my friends in the environmental movement. How can we care so deeply about wild animals, beautiful landscapes, clean cities, and the sane use of chemicals, yet see the world so differently? I could just as well ask why all the polls show that Americans overwhelmingly want a better natural environment, despite signs we've made little progress. The reason is that as the environmental movement has gathered strength, its vision has become ever more fixed on vengeance and utopia. My friends in the movement will protest, but they speak and act as if business, industry, technology, and even science are villains to be arrested and punished. Their ideals include pristine wilderness, organic farming, redistributing land and income, living as coequals with nature, herbal medicines, and generally doing with fewer of life's pleasures.

Sophisticated environmentalists dress up the longing for vengeance and utopia in layers of rhetoric. As much as they talk about diversity of opinions and solutions, however, they are addicted to grief for some lost world.

Instead of talking realistically about the future, they substitute vague notions of recapturing a golden past. Like the lodestone that moves every compass needle on earth, these forces lie beneath the surface. From radical ecosaboteurs to the White House and Congress, environmentalists are facing the wrong direction.

In 1992, when Bill Clinton and Al Gore won the White House, the executive director of the Sierra Club, Carl Pope, spoke for his 575,000 members and most other environmentalists when he said, "We look forward now with hope for a new environmental knowledge and understanding by our national leaders, and forward movement for environmental protections."[1] No administration has ever been in fuller sympathy with the environmental movement. Vice President Gore's book, *Earth in the Balance*, lays out his views and feelings, in which the pull of the utopian is obvious. Environmental problems are not mistakes; instead, they reveal flaws in our basic values. The industrial world is a "dysfunctional family" divorced from nature. It is the same voice and the same vision I hear from my friends and read in their books. The vice president's solutions, from the general to the specific, are also mainstream environmentalism: a call for massive government expenditures, sweeping regulations, government choosing to encourage or discourage specific technologies, and the elimination of the internal combustion engine.

Clinton and Gore immediately began picking people from the movement to run environmental programs. The White House may argue with the movement over details, but solid agreement exists on the basic direction—more and stronger regulations, wider government authority over land use, more government control over industry and technology. The political process inevitably dilutes ideals, but this administration has made it clear that it wants to accelerate and expand all of the past approaches to environmental problems.

The Reagan and Bush administrations, which were dependent on a Democratic Congress, did not change the course of environmental policy but a few degrees from the Carter years. In fact, environmental regulations and our spending on them mushroomed. So did enforcement of environmental laws. The movement survived all challenges to its hold on public thinking and government programs. With the election of President Clinton and Vice President Gore, the movement has begun to press with new vigor its old ideas.

The environmental movement is quickly charting a new way for American thinking. Instead of objective science, it promotes an adversarial science. Instead of seeing the natural world as a manageable system, it sees a fragile, unmendable lace of life. Instead of trying to improve our ability to manage natural processes, it believes we should keep our hands off whenever possible. Instead of individual responsibility for managing, it wants centralized government management. Instead of recognizing that economic growth goes hand in hand with environmental stewardship, it has declared growth the enemy.

We have improved water and air quality and saved some important wild lands, but almost always with unnecessary expense and divisive political battles. If we accept the environmental movement's own assessment of the world, the record is not good. After twenty years of intense activity, the movement should be judged by its record. As a long-time activist for environmental improvement, I will be called a traitor to the cause for my opinions, but it has become clear to me that either we care about the environmental movement or we care about the environment.

THE QUESTION EVERY ENVIRONMENTALIST SHOULD ASK

As the movement gets stronger, why does the quality of the environment get worse? This doesn't prove cause and effect, but it says the environmental movement so far, considered overall, has failed.

The Worldwatch Institute, Audubon Society, and National Wildlife Federation all publish report cards on environmental trends. Their indexes of environmental quality continue to show a consistent decline. We pass a Superfund Bill to clean up toxic waste, but ground water contamination grows. We have an Endangered Species Act, but it saves no species. We have a Clean Water Act, but water quality continues to deteriorate. We have a wetlands protection program, but we continue to lose wetlands. We have energy conservation programs, but we consume more energy. At least that is the picture environmental groups paint.

A 1993 issue of *Audubon* depicts the environmental basis of world hunger with maps and charts that show hunger spreading, more soil erosion, more deforestation, accelerated global warming, and ocean life dying from radiation streaming through the ozone "hole."[2] If the indexes are wrong, it suggests that the environmental groups that compile them are ignorant of science or, worse, that they are lying. If the indexes are right, the only way to justify continuing on the present path is to say we just do not have enough regulation, government planning, and social control.

Yet in the 1980s and early 1990s regulations multiplied. So did enforcement, although anti-Reagan environmentalists did not publicize it. In 1982, the U.S. Justice Department formed its Environmental Crimes Section. The U.S. Environmental Protection Agency (EPA) added criminal investigators to its technical staff. By the end of the Reagan years, the feds were sending environmental criminals to jail. In 1989, federal prosecutors won 107 guilty pleas, up 70 percent from 1988. Fines totaled $13 million, and corporate officers received sentences totaling thirty-seven years in jail. In 1992, EPA, working with North Carolina officials, won $5.2 million in fines from just five companies. President Bush's administration nourished environmental regulations until they became 38 percent of the regulatory budget. Yet according to environmental groups, these were disaster years for environmental quality. Greed and polluters controlled government. Almost everything got worse.

Environmental groups believe their indexes. Their solutions assume that it is necessary to speed up in the same direction they have been moving—increase regulation and government control, consume less, slow down technology, redirect scientific thinking. Above all, lead simpler lives—buy less, drive less, eat little or no meat, live colder in the winter and hotter in the summer, have fewer children, decrease competition, and farm organically.

WHAT'S WRONG?

Clearly something is wrong. I see three major possibilities why:

1) We have not gone far enough with the thinking and methods that environmental groups embrace, and when we go farther down this path, we will begin to see things change for the better.
2) Environmental groups do not understand the problems. They rely on bad science.
3) The way environmental groups measure the problems and construct solutions is making things worse.

If we just have not gone far enough with regulations and trying to change society, I am not sure we want to risk going much farther, given the correlation between the growing power of environmental groups and the continued decline in nature's health (assuming the environmental indexes are correct). If we have not done enough to change people's thinking, I worry that further efforts will become dictatorial. Attitudes change slowly, and the movement toward political correctness has already demonstrated that policing thought quickly becomes undemocratic.

The United States and industrialized countries in general do care about nature, and I will show that we care less selfishly than any other culture. Our thinking is not at fault, and our science has positioned us to know nature better than any human beings have ever done. In short, it would be wrong to sacrifice democracy for a clean environment and to assume that the answer lies in some culture outside industrial nations.

Concerning the second possibility, several recent books have confronted the science that environmentalists use to support their solutions to problems.[3] The confrontations seldom decide who is right, but they have done two other things. First, they have exposed many misconceptions about global warming, synthetic chemicals, agriculture's Green Revolution, endangered species, rain forests, wetlands, acid rain, and deforestation. Second, they have demonstrated abuse and outright fraud in the way environmental groups have manipulated scientific research. Nowhere has this been clearer than in the measurements they have made of environmental quality. Time and again the environmental movement has chosen to measure the quality of life and of the environment by taking their statistics from highly selective segments of time in which the trend seems to be getting worse. In the 1950s it chose temperature measurements that showed the earth cooling off. In the 1980s, when American summers and winters got warmer, global warming was declared a fact by the movement and became the symbol of destruction for environmental writers, including the *New Yorker's* Bill McKibben and Vice President Al Gore. As this book goes to press North America, Europe, and Central Asia have all suffered record cold weather. Real science has not yet detected a clear measure of either global warming or cooling.

Since 1970, the environmental movement has been producing figures to show that the world will soon run short of food. In January 1994, the influential Washington-based World Watch Institute got global coverage for its annual "State of the World" report, which said figures showed population had begun to outstrip food production and technology's ability to feed people. Keith Collins, acting assistant secretary of agriculture, pointed out that, to the contrary, grain production has been rising 3 percent a year and population, only 2 percent. He said World Watch had used figures from the 1980s when the rate of increase in production slowed because of an overabundance of food and decreasing prices. He pointed out that even in inflation-adjusted dollars world-wide food commodity prices had decreased 75 percent since the 1950s. World Watch's media star, Lester Brown, claimed that we had run out of new technologies to increase production. Collins pointed out that the 1980s had just begun to apply the vast potential of genetic engineering.

Refereeing scientific arguments is the business of scientific journals, but the public cannot always wait. Besides, if we waited for absolute certainty from science, we would wait forever. So you won't find the scientifically proven answer to any hot environmental question in this book. Great environmental problems have no final solutions—except in congressional hearings, at environmental rallies, on talk shows, or in barber shops. To say we know what nature should be or exactly what the future must bring is as silly as planning a house that will be built thirty years from now. Consider this: thirty years ago, typical homes were not wired for computers, did not have heat pumps, often had asbestos pipe insulation, used huge amounts of solid wood, and had single-pane windows. My concern with the findings of science is to propose how we can use them intelligently and honorably.

If the cause of nature's ills is not too little environmentalism or too much bad science, can it be that environmentalism itself is the problem? That is the unlikely argument of this experienced environmentalist.

ENVIRONMENTALISM AND THE POWER THAT DIVIDES

After thirty years in the environmental movement, I am worried that as it gains power, it cares less and less about reason and science. Its influence on movies, academia, and literature has already turned history into fiction and propaganda. The purpose of rewriting history is to influence the present and control the future. In short, I believe the environmental movement has almost lost touch with reality. Through its recent political successes, it has started to exercise power in ways that may do more harm to nature than good.

Over the past twenty years the environmental meetings I attend as a journalist have seemed more and more like church meetings. Every year for five years I have sat with a group of widely published nature writers, most of them also respected lecturers and professors. These writers compile the anthologies that schools and universities use in courses on literature and the environment. This is a little like having a single religious sect write student textbooks. In our meetings, they talk about "converting"

America and not letting science interfere with a "spiritual understanding" of nature. They talk about all creatures being "sacred." Perhaps we cannot prove the value of every species, they say, but we must make a "leap of faith." This pulpit rhetoric has become the standard in environmental writing and teaching.

A certain angry and violent language accompanies the religious language. Three days after the election of Bill Clinton and Al Gore, the nation's oldest environmental group, American Forests (founded in 1875), organized a meeting, "People as Positive Agents of Environmental Change." Forester Mollie Beatty, later appointed to head the Fish and Wildlife Service in the Clinton administration, talked about people outside the movement as "Joe and Jane Sixpack." Another speaker said that the only way to get the public's attention is to "hit them with a two-by-four."

No one expects to see environmentalists in the streets with two-by-fours, sort of a fascist Green Shirt movement, but their language is not the language of reason and it does not invite democratic debate. As the 1980s ended with too little progress on too many problems, the movement's apocalyptic tendency took over. Nature is as good as dead, declared the *New Yorker* writer Bill McKibben in *The End of Nature*.

Environmentalists everywhere insist that the 1990s are the last chance to save the earth. Their fears were made official in Vice President Gore's book, *Earth in the Balance*. The crisis mentality demands large-scale and extreme action. Environmentalists are espousing a conservative dictum pronounced by Barry Goldwater in the 1960s—extremism in the pursuit of virtue is no vice.

Like preachers in the pulpit, environmental leaders rage against the "seduction" of our society and of struggling poor countries by consumerism. It is as if people and countries have no free will, or worse yet, as if their freedom of choice should be taken away. Take away free choice and society always ends up with official choices dictated by an elite. Diversity of choice, like diversity in nature, is the key to civilization's survival.

An old movie plot involves the hero who is waylaid in a lonely spot by the archvillain, who then puts on the hero's clothes and disguises his face. The destroyer goes forth dressed as an angel or the Good Samaritan. I will show that something like this happened when today's environmental movement shoved aside a conservation movement started by forward-thinking men and women in the nineteenth century.

Of course, the environmental movement promises to do good, and for most the intent is sincere. In fact, many environmentalists believe in their own goodness too passionately, like the small-town minister who was feuding with a nearby priest. One day the two met on the street. The priest said, "My friend, isn't it time we tried to work together? After all, we each serve the same God."

"Yes, we do," the minister replied. "You in your way and I in His way."

Such is the division between America's environmental activists, who claim to know how nature wants us to behave, and the rest of the country, which wants to behave correctly toward nature but is not necessarily sure how to do so.

Although no one may love every aspect of nature, we all love it in general. How do we love it? As the poet said, "Let me count the ways." Look on any living room wall and you will probably find flowers, or a landscape or seascape. Read the personal ads, where our quest for love is so often connected to the outdoors—"nature lover," "quiet walks in the woods," "the beach," "a fire"—nature is part of our idea of the best in life. Browse through the literature at a travel agency and almost every glossy brochure features one or more of the following: snow-capped mountains, blue waters, white sands, glaciers, trees. Survey the greeting card racks. Name the most popular children's movies and the most romantic adult movies. Nature is everywhere and every American loves it. This unites us, but what divides us will become increasingly important to the world's future. This book is about those things that divide us and about the fundamental values that should and could unite us.

Differences in how we understand nature and what we should do about environmental problems have created a major split in our political life. Environmentalists scorned George Bush when he said he wanted to be the environmental president, and they scoffed at the idea that Ronald Reagan had even a passing interest in the welfare of nature. But now the power has shifted. Vice President Gore has said that protecting the environment must be the "central organizing principle" of society and of his work in politics. For him that means

> embarking on an all-out effort to use every policy and program, every law and institution, every treaty and alliance, every tactic and strategy, every plan and course of action. . . . Minor shifts in policy, marginal adjustments in ongoing programs, moderate improvements in laws and regulations, rhetoric offered in lieu of genuine change—these are all forms of appeasement, designed to satisfy the public's desire to believe that sacrifice, struggle, and a wrenching transformation of society will not be necessary.[4]

The vice president is not specific about what that "wrenching transformation" might be, but he leaves little doubt that his view of the world is the environmentalist's view. His prescriptions for solving problems urge us farther down the path we have been traveling for more than twenty years.

Although I have served several terms as president of two statewide environmental groups, the Conservation Council of North Carolina and the Conservation Foundation of North Carolina, I now worry that the triumph of the environmental movement is about to infect our thinking with a wrong-headedness that will require precious decades to reverse. Someone once remarked that uninformed activists specialize in ten-second critiques that take a half hour to rebut. Unfortunately, when the public and Congress try to understand dozens of complex issues such as health care, crime, global trade, and international terrorism, simple ideas and sound bites often dominate the political process, and when simplistic thinking becomes law and regulation, the mistake may take years to remove from the books.

The economic dangers of simple-mindedness are easy to measure. Take proposals based on the fear that carbon dioxide from cars, factories, and power plants has caused the earth to enter a disastrous warming period; a big bandwagon gained momentum before anyone gave it adequate steering or brakes. Al Gore has declared that global warming is a fact and an immediate crisis, "an environmental holocaust without precedent."[5] This is the kind of language that demands immediate and dramatic action. To stabilize U.S. carbon dioxide emissions would cost $95 billion in the first year, according to a study by Charles River Associates.[6] If we are to spend such huge sums, we had better know what we are doing. We don't. Despite certainty in the environmental community, there is none among scientists who work directly on the problem. Even those who predict some warming cannot determine how much it might be or what the best solution is.

Another example of how simple-minded thinking can lead to expensive mistakes is the question of toxic and cancer-causing chemicals. Few public campaigns have cost more money than those against toxic chemicals. However, Dr. Bruce Ames, a biochemist and inventor of many tests for cancer-causing chemicals, has said, "When looking at causes of cancer . . . pollution is almost irrelevant."[7] Many scientists disagree, but Ames's tests for carcinogens are still respected and used.

E. Donald Elliott, a Yale law professor and former EPA general counsel, said, "I have never seen a single rule where we weren't paying at least

$100 million per life [saved] for some portion of the rule."[8] Some rules, he said, cost more than $30 billion. EPA rules passed in 1990 about wood preservatives have cost $5.7 trillion for each life they were estimated to save, according to John Goodman of the National Center for Policy Analysis in Dallas.[9]

The Endangered Species Act has not saved any species, but the costs for trying have run into the billions. In Travis County, Texas, for example, listing just two songbirds as endangered caused a $300 million loss in property values by limiting development potential, according to the deputy chief tax appraiser.[10] Perhaps it is worth it, but we should know we are playing this game for high stakes, especially when we extend this approach to international treaties, as we did at the 1992 Earth Summit in Rio de Janeiro. In the debate over extinctions, *Science*, the magazine of the American Association for the Advancement of Science, pointed to a series of predictions of which only four of twenty-two "came with sufficient explanation to permit independent examination." A leading conservationist, who would not give his name for fear of retaliation from his colleagues, told *Science*, "A gut feeling isn't much backup when you're asking people all over the world to change their lives completely."[11]

Millions of dollars are spent excavating, treating, and reburying soil contaminated by leaking fuel tanks from service stations. In most cases, the hydrocarbon chemicals in such soil would break down into their simple and harmless components if spread in the air and sun, and tilled or composted, but all cases are treated as expensive worst cases.

The well-intended Superfund for toxic waste sites has been another expensive mistake. Of the $10 billion spent on Superfund cleanup between 1980 and 1991, the congressional Office of Technology Assessment found that researchers and administrators, not cleanup crews, received $6 billion.

NATURE HAS NOTHING TO SAY

From Vice President Gore to environmental fringe groups, the movement justifies these expensive diversions and failures by saying that we are in a crisis: the 1990s are supposed to be a "make or break" decade for the earth. I agree, but with one exception. First, the crisis is not for the earth. The earth is not "in the balance." The quality of our life and the character of our environment are in the balance. The earth and nature will long survive us. This is an important distinction, because environmentalists often base their actions on what they say nature wants, needs, or prefers. In fact, it is humans who set preferences for nature. Mother Nature is mute, and environmentalists have rushed to put words in her mouth.

The words are clear and often persuasive. But it is a script that plays well only in a small corner of the global village. Human preferences have changed radically over short periods of history. Even today, a peasant in Guatemala, an African herdsman, an Indonesian tree farmer, a Haitian refugee, and a New York socialite do not prefer the same nature that environmentalists prefer. As for nature, it has exhibited many different states over the four billion years earth has existed, and its only constant preference has been change. Anyone who knows earlier periods of natural history cannot possibly conclude that the long periods of conditions hostile to human life indicate that nature prefers our present mild environment. In the familiar comparison of earth's history to one week on the clock, neither human life nor nature as we know it counts for much:

Monday, one minute past midnight: earth is born
Saturday, 4 P.M.: dinosaurs appear
Saturday, 30 seconds before midnight: modern humans appear
Saturday, 1.5 seconds before midnight: agriculture begins

Nature's changes roiled across the surface of the earth for almost 6 billion years before mammals appeared, 70 million years ago. The Proterozoic period favored an oceanic soup of algae and bacteria, worms and sponges. The Devonian was a time of snails, fish, coral, and a few amphibians. The Jurassic—135 million years ago—favored dinosaurs and fernlike plants. Forests as we know them only began to grow worldwide with the appearance of large mammal populations. Human beings and their nature are a very young experiment. The biologist Garrett Hardin suggests we have little reason to think humankind will last any longer than dinosaurs or other bygone creations.[12] Anyone who thinks that nature prefers humans and our environment or any species we know and love should consider that 99.9 percent of the billions of species that have lived on earth over the past 3.5 billion years have been dismissed into oblivion.

Nature does not prefer the conditions we love, or us. To the contrary. She does not prefer nature as we know it. We do and we should. But to preserve it for a world of six billion people means taking nature's fate in our hands as never before. McKibben denies this in *The End of Nature;* the book is a desperate plea for "hands off nature." However, if we are to survive with our preferred natural support system intact, we must put our hands on—quickly and intelligently.

Environmentalists have never forgiven the Bible for announcing in chapter 1 of Genesis the purpose of God's creation. Having made the world, God tells Adam and Eve: "Be fruitful and multiply, and fill the earth and subdue it; and have dominion over the fish of the sea and over the birds of the air and over every living thing that moves upon the earth." Perhaps if the people who wrote the Bible and recorded the privilege of dominion over the land had any special intuition about nature, they knew that rising above the brutal and precarious life of nomads warring for land and resources required the conquest of nature's preference for change and making nature's resources available for human improvement. Thus, the passage that environmentalists fear and condemn takes on a new light. This passage recognizes that if we want to make nature a comfortable place and limit its changes to our tolerance levels, we had better develop the tools to manage it.

My reading, of course, is a historical view and does not affirm or deny a divine hand guiding the pen. Theologians would say this passage assigns humankind a position of command between God and nature; they may debate the purpose of such a position, but the results are the same no matter why God put us there.

Environmentalists who say they know what nature is and what nature wants are supreme egotists. When we look at the difference between what modern ecology reveals about nature and the Garden of Eden that environmentalists prefer, it becomes clear just how far this fantasy goes in preventing us from exercising intelligent management over the real world.

The basic assumption of this book is that nature does not care—only we do. And because we care, we have made the world an ever more livable place for ourselves by using science and technology and by exercising our power of dominion, wherever it comes from. The exercise of this power is a natural act. It is rational. We are doing what every animal in creation has tried to do, only we are doing it more successfully because we have imagination and technology. A curious fact serves as a quick measure of how great our success has been. Almost all mammals live for about 200 million breaths and 800 million heartbeats. A mouse uses up these measures and dies much faster than an elephant because the lungs and the heart work fast. Large mammals live longer and their hearts and lungs work more slowly. On average, civilized humans exceed their allotted breaths and heartbeats by some fifty years. Humankind's success, and how to build on it, is this book's compass. In the process of building our cultural endowment, we have made mistakes that have not been kind either to nature or to our own species. In general, however, for each of the giant steps forward, we have taken only a few smaller steps backward. Our progress has been the result largely of Western science and technology. Unlike cultures that only feared and revered nature, industrialized cultures have pursued dominion over nature and subdued most of its dangerous tendencies, achieving what no other culture has done. No other tradition has developed a sophisticated technology capable of feeding six billion people and monitoring the condition of the environment. To argue that Western culture is responsible for the necessity of caring for so many people is irrelevant. There is no turning back. We can find beauty, knowledge, and insight in the cultures of Bushmen and Buddhists, but so far only Western science and technology continue to provide for the material needs of the world's people. A movement that rejects this tradition is dangerously out of touch with reality, and cavalierly robs less fortunate people of their best hope:

> We are led to forget the dominating misery of other times in part by the grace of literature, poetry, romance, and legend, which celebrate those who lived well and forget those who lived in the silence of poverty. The eras of misery have been mythologized and may even be remembered as golden ages of pastoral simplicity. They were not.[13]

The questions of this book are how do we intelligently determine the real state of our environment, decide what tools we need, and manage for the long-term success of the human species? We cannot answer these questions if we continue to insist that our culture has failed because four-fifths of the world still lives in relative poverty. History has been full of pain, but we can look at the long sweep of 20,000 years past and see that more and more people are being included in the benefits of technology, especially in the past one hundred years. Most demographers agree that the world population will level off in the next one hundred years. Meanwhile, technology continues to broaden our understanding of the world and multiply our options for preserving the full diversity of nature that supports us. A new Renaissance in human understanding has already begun. We must distinguish between the fear and trembling of ignorance and the prudence and courage that are among humanity's most valuable traits.

Our heritage has one single strength that atones for all of its weaknesses, that gives us an opportunity to save ourselves and nature, and that more than outweighs our tolerance for destruction. Western-style democracy, with its ever expanding protection of individual freedom, offers invaluable opportunities and incentives to pursue creative answers to environmental problems. Despite constant argument, both conservatives and liberals are busily using these opportunities.

Both advances in science and the leisure brought by our economy have given us a greater chance than ever to know nature as it really is. Science seems to be marching forward and the environmental movement backward. The environmental movement is lively, eloquent, and entertaining. Science is full of people doing tedious work, reading through highly specialized reams of data. Here is where our materialism does threaten to blind us. We pay attention to science mainly when it produces some useful new product.

Like the people who doubted Columbus, environmentalists stand on the shore and talk of the folly and arrogance of mastering nature. They may be right. But they are too late. When I was teaching English in the late 1960s, a colleague wondered why I was in Washington, D.C., lobbying to save wilderness areas. Wilderness does not exist, he said, because by drawing a line around it, we have made it unnatural. All the world has become our garden. Whether or not we like it, we have become the managers of the natural world. We have the means to destroy life as we know it. We can determine the composition of the atmosphere and the levels of the oceans. If nature's survival is at stake, so is our own. With a population of six billion and growing, we must manage nature for its sake and ours. There is no turning back.

CHAPTER 2

The Search for Authority

Groups cannot organize themselves without a purpose. Within a society, culture, or nation, power comes to the group whose story of history can persuade, or be imposed on, large numbers of people. It makes little difference if the story comes from shamans, astrologers, priests, philosophers, or environmentalists. – or megacorporations, indust/military complex.

To answer the questions how and why the world and humankind were created, every society, large and small, has told a story. Once those questions are answered, the story turns to how things go wrong when someone forgets the purpose of creation or tries to change the rules. In primitive societies, myth and history are often the same story. In our society, scripture and history overlap, but in large part, history begins where scripture ends and has a power of its own.

Environmentalists have used history to legitimize themselves and to support their vision of the world and our purpose in it. Since their view of natural history is shaped by the movement's own history, that is where I begin.

Environmentalists act as if their movement has been endowed by history with the right to be the voice and conscience of science. Environmentalism is selective in its use of history and science to influence and change the political system. Other movements act the same way—the right to life movement, the American Civil Liberties Union, and the National Rifle Association, to name a few. Among such movements, the environmental movement relies most heavily on the natural sciences.

The environmental movement has appointed itself as the public voice of science. But is environmentalism any more scientific than the "scientific communism" millions in the Soviet bloc had to study in high school, or "Christian Science," or "Scientology"?

In a diverse and democratic society, most movements like to claim scientific backing, if not proof, for the correctness of their political prescriptions. The phrase "it's scientifically proven" is a hundred times more common in American political debate than "it's God's will." Yet Americans have only a passing interest in the difference between science and faith. Because the environmental movement uses the vocabulary of science, has scientists among its most vocal members, and often claims overwhelming scientific consensus for its positions, it declares that history has made it the voice of science.

When a scientist such as the butterfly specialist Paul Ehrlich stakes out a public position as the environmental spokesman on population and economics, many people assume that what he says is based on scientific research. When the popular astronomer Carl Sagan asserted that Kuwait's burning oil wells would create devastating climatic changes, his reputation as a scientist cloaked his woefully inaccurate and unresearched predictions in the aura of objectivity. In his case the aura is enhanced by his boyish sincerity and his slightly otherworldly gaze.

> Now it is a characteristic of such intellectuals that they see no incongruity in moving from their own discipline, where they are acknowledged masters, to public affairs, where they might be supposed to have no more right to a hearing than anyone else. Indeed they always claim that their special knowledge gives them valuable insights.[1]

The media generally do not distinguish between scientists speaking as scientists and scientists speaking as part of the environmental congregation. A Ph.D. in a science somehow entitles the holder to be taken seriously about any topic, unlike a Ph.D. in English or music.

> Advocacy disguised as science is this age's most powerful means of persuasion. But most people are unaware that many scientists have subordinated their science to political ends, so they continue to regard science as being objective "Truth." The well-misinformed public has been suckered into concentrating authority into the hands of a "knowing elite." The earth is said to be in crisis. But it is the growth of scientific advocacy—be it concerning acid rain, global warming, or any number of other supposed "catastrophes waiting to happen"—that is the real crisis.[2]

[handwritten margin notes: not all knowledge is lettered. How many of our political leaders have any credentials or real advice from experts? MARKETING *and* Hard today to distinguish who to believe about any thing.*]*

Ecology, the science, and environmentalism, the movement, are very different creatures with very different historical roots. Since the subject matter of both is nature, their vocabulary and interests often overlap. But then, so do the subject matter and interests of capitalism and communism. Perhaps science and environmentalism are not such polar opposites, but they are far from being twins. At heart, they are two different ways of seeing and thinking about our place in the universe. Science is *a method* of knowing the world. Environmentalism is *an ideology* that tells us what to find. Science evolved from ancient Greek philosophy as a way to explore the natural world. Environmentalism began to evolve 1,600 years later as a reaction to the new power given to science when wealthy Renaissance businessmen and nobility wedded it to technology.

Deliberately or not, both friends and foes of the environmental movement confused history in the 1970s by calling environmentalists "ecologists," despite the fact that many are not scientists. When the Sierra Club published Father Thomas Berry's *The Dream of the Earth* in 1988, the Jesuit priest wrote of his new theology as "the ecological vision." The famous center for New Age training, New York's Omega Institute, advertises the scientific degrees of teachers of several of its courses, yet these courses throw out scientific method and substitute subjective and unverifiable insight. Scientists themselves have been seduced by environmentalism, if not by the logic of the movement then by its claims to virtue. Krug says, "We scientists, who are trained skeptics, automatically and unthinkingly assumed that people associated with the 'just cause' of the environment are themselves just."[3]

A NEW EGG IN AN OLD NEST

In the 1960s, when I first became active in the environmental movement as a member and later president of the Conservation Council of North Carolina, we often used the words *conservation* and *environmental* interchangeably. Yet we knew the difference. One of our earliest questions in building the organization was how to involve more people from "conservation groups." We meant people whose favorite outdoor activities were hunting and fishing and whose chief goal was to manipulate nature scientifically.

In writing their own version of history, environmentalists have minimized the role of conservationists. When the environmental movement first began reaching for political power and control of the public imagination, it was careful to give some credit to the conservation movement, which was dominated by sportsmen and sportswomen. In fact, the environmental movement thought of itself, or tried to make others think of it, as continuing the tradition inspired by President Theodore Roosevelt, John James Audubon, George Bird Grinnell, John Burroughs, and Aldo Leopold—all unabashed hunting enthusiasts.

In 1970, when the Sierra Club published *Ecotactics: The Sierra Club Handbook for Environment Activists,* it included an essay by the club's director Paul Brooks, "Notes on the Conservation Revolution." Brooks was already hard at work trying to change the conservation movement into environmentalism. Brooks wrote that Americans should not think that the values of the conservation movement are accepted; doing so "conceals the fact that the conservation movement, though it operates within the law, is in principle revolutionary."[4]

The conservation and environmental movements never really merged. In the 1960s, seventy years after its first organizations were formed, the conservation movement began to lose ground. No president since Teddy Roosevelt had been a dedicated outdoorsman or energetic champion of preserving wildlife. The environmental movement simply took over because the time was ripe. Many conservationists participated in both movements. Some found in the environmental movement the true religion they had always hoped for but never imagined. Others went along because the environmental movement promised more of what they enjoyed— wilderness, wildlife, clear waters, open beaches. What's more, no one would have to pay. Government would recognize our rights and give everything to us.

The environmental movement likes to think that Americans were awakened by men such as the Sierra Club founder and writer John Muir. On a practical level, Muir's influence was not absorbed by activists until the environmental movement took off in the 1970s. In the nineteenth century, when Muir was still considered an eccentric young dabbler in the High Sierras, other Americans set to work reversing the bad management practices of the country's loggers, miners, ranchers, and farmers. A hundred years before the first Earth Day, the conservation movement began as the "club movement." The clubs were formed after the Civil War by men who hunted and fished for recreation. By 1875, the movement included one hundred sportsmen's groups. By 1880, the number had tripled and included clubs from New York to California.

Many of the conservationists whose biographies have been selectively edited to suit the environmentalist version of history loved hunting, fishing, and killing animals. Here's a quick list:

John Burroughs, bird watcher and author: fishing

George Vest and Samuel Cox, congressional advocates of Yellowstone National Park: hunting and fishing

George Catlin, painter of American Indians and the Great Plains, the first to propose a national park: buffalo hunting for sport

John James Audubon, painter and ornithologist: bird hunting

C. Hart Merriam, founder of the U.S. Fish and Wildlife Service: hunting

Frederick Law Olmsted, park designer: fox and whooping crane hunting

William Dutcher, founder of the Audubon Society: bird hunting

Theodore Roosevelt, first conservation president: buffalo and African big game hunting

Frank Chapman, ornithologist: bird hunting

The founding of the conservation movement by sportsmen endowed it with a scientific and practical mind, although it did not lack for feeling and sentimentality. In addition to saving game animal habitats and fishing waters, they lobbied for the first national and state parks and forests. Issues that today seem to have been brought to the public by the environmental movement were made hot topics a century ago by sportsmen. They protested industrial chemicals and mining waste that polluted streams and lakes. They attacked commercial net fishing and clear-cutting of watersheds. They sponsored scientific research on fish breeding and restocking. Sportsmen began tagging fish to trace their life cycles as early as 1896.

The early conservationists understood the importance of preserving an entire habitat. Fishermen lobbied not only for hatchery-raised fish to be stocked for their catch, but also for maintaining stream and river bank cover. Hunting groups often bought or leased game preserves where nature ran its course. The many subdivisions and towns named Deer Park are ironic tributes to these preserves, often called deer parks. As early as 1871, the Blooming Grove Park Association established a 12,000-acre preserve with a 1,000-acre deer park in Pike County, Pennsylvania. (It is worth noting that private landowners not only established this park a year before Yellowstone National Park, but they also had an overall management plan, which Yellowstone and other government preserves would not create for more than a decade.) Even the country's largest state park, New York's Adirondack Park, was begun by conservationists organized as the Bisby

Club and the Adirondack League Club. When the two clubs merged in the early 1890s, they controlled almost 190,000 acres.

Sportsmen also succeeded in passing the country's first widespread environmental laws, often over the objections of small family farmers, who have now become the model of old rural virtues for today's environmentalists.

Ecofeminists should note that nineteenth-century sportsmen (with help from like-minded women) put an end to the craze for ladies' hats adorned with exotic feathers and even whole birds. The rarer the pretty birds became, the higher the price went. Congressman John Lacey, whose 1900 Lacey Act ended all market hunting and interstate trade in wildlife products, was a dedicated hunter and fisherman.

The high point of the early conservation movement was the election of the only person so far who can legitimately claim to have been an environmental president. Hunter and explorer Theodore Roosevelt learned the science of conservation from his friends and advisers Gifford Pinchot, a forester, and George Bird Grinnell, editor of *Field and Stream.* Roosevelt added 100 million acres to the country's forest preserves, started a professional national forestry service run by scientists, established five national parks, and set aside more than fifty wildlife refuges.

Despite a great overlap of feeling and political agreement between conservationists and environmentalists, their disagreements may be more important. The most telling area is hunting. Environmental groups would oppose hunting with vigor, except they know they would lose important allies. Instead, they try to give the question a wide berth, like not talking about toilets at the dinner table. We all need toilets, and we all have one or more in every house, but we do not talk about them at the table. So, too, all animals are killed, usually by other animals, and we can talk about that, but not about the human animal killing animals, whether it is a hunter or a butcher. Vegetarian environmentalists try not to think about the chickens, four to a cage, who lay the eggs for omelets and quiche. Or about the size of the oak tree that a logger felled to make their book shelves and kitchen cabinets. But just as members of even the most puritanical religions engage in sex, most environmentalists engage in killing animals and felling hundred-year-old trees. Like all puritans they prefer to speak of such acts as something other people do, or as a generalized sin against which we all must struggle.

Environmentalists do not believe that human beings should destroy any part of nature, except under extreme circumstances. They think of animals as individuals, not as populations. For them, the interdependence of humans and nature means equality. This is not just humility. It is a denial of human nature. By placing humans on the same level with the rest of creation, environmentalists have dehumanized us.

In public policy, environmentalists dictate that natural resources should be handed on to the next generation intact. We can do this with a park or a wilderness area, but we can hardly do it with minerals or fossil fuels. To pass these resources on intact or undiminished would mean to use none now. And future generations, under the same ethic, could also use none. It comes very close to a "stop-the-world-I-want-to-get-off" formula. Environmentalists propose it, but none live by it.

Actress Meryl Streep, for instance, became an antitoxics crusader for the environmental movement. The pesticides she is campaigning against are necessary to provide food to a growing world population. Since environmentalists frequently testify that affluent children consume five times as much as poor children, they might want to ask Streep why she had that fourth child.

Streep and other environmentalists, of course, continue to preach guilt and gloom about our own species. Even if they acknowledge, however quietly, the impossibility of achieving the equality of all species—of destroying nothing, changing little—the ideal nevertheless pulls the movement in that direction. Streep's mentors in the Natural Resource Defense Council and the "experts" from World Watch Institute have been very active advisers to the governments of developing countries. Whether they live by their own version of history or not, they are pushing public policy around the world in that direction.

When environmentalists reinvent history to legitimize their perspective, they are taking control of the public memory. Every totalitarian leader has known the power of controlling history; the historian C. Vann Woodward wrote that the person "who controls the past controls the future."[5] If history condemns something in the past, the public is more likely to try to ban it in the future.

Environmental leaders, intent on writing history their way, admit a poor grasp of history as written by others. In a survey of several hundred environmental groups conducted in the late 1980s by the Conservation Fund, only 13 percent of the environmental leaders rated newly hired staff as having an excellent knowledge of conservation history. Only 18 percent thought having discussions with leading thinkers in conservation history would be very useful.[6]

The eighteenth century Philosophers, like the medieval scholastics held fast to a revealed body of knowledge, and they were unwilling or unable to learn anything from history which could not, by some ingenious trick played on the dead, be reconciled with their faith.[7]

The biggest difference between the conservation movement and the environmental movement is larger than any single issue. It is a philosophical difference from which all other differences arise. Conservationists believe that humankind can manage nature through the intelligent use of science. Environmentalists believe we must back off and let nature manage itself. It is the difference between optimism and pessimism.

This difference also represents conflicting views of human nature. Conservationists believe humans are capable not only of managing nature, but also of changing their management techniques to suit new circumstances. Environmentalists believe this is arrogant and tragic. Nothing demonstrates their attitude better than their use of statistics. When they predict global warming, overpopulation, and wildlife extinctions, they usually assume that any current trend will continue at the same rate: accelerating curves continue to accelerate, and straight lines cannot be bent. Except by catastrophe. In contrast, I believe that

> Nothing ever changes at the same rate indefinitely. This is why babies do not grow up to be the size of barns; why the surface of the earth is not covered with a meters-thick layer of the progeny of a single bacterium (though it is covered with thick academics); and why the fellow who put a shilling into his savings account 200 years ago does not now own all the money in the world. When the monomaniacal application of a rule leads to ridiculous results, one should consider that the rule or its application are in error, not straightaway conclude that the world is wrong.[8]

THE PROPHETS AREN'T ALWAYS RIGHT

When talking to environmentalists I find they allow only two fates for the earth: the world has just ended or is about to end. Everyone is welcome to talk about the good old days, but no one is supposed to ask when they were. And you must speak with proper skepticism about the possibility of survival. Survival must seem as unlikely as the Kentucky Derby being won by a three-legged horse. Environmentalists generally don't laugh much at themselves for two reasons. First, like most true believers, they do not find themselves funny. Second, it is not polite to laugh at doomed people. Humor among environmentalists is generally gallows humor. It never occurs to them that they may be false prophets. Yet men and women have predicted environmental catastrophe over and over throughout the years—and been wrong:

1865: COAL SUPPLIES TO RUN OUT

In 1865, the English economist and social critic William Stanley Jevons predicted that England's coal supplies would be used up quickly by the expanding population of the prosperous Victorian era: "The conclusion is inevitable that our present happy progressive condition is a thing of limited duration."[9]

1886: NO MORE BIRDS

"We may read [the story] plainly enough in the silent hedges, once vocal with the morning songs of birds, and in the deserted fields where once bright plumage flashed in the sunlight."[10]

1887: HUMANS WILL CONSUME ALL OF NATURE

"There will soon not be a bird of paradise on earth, and the ostrich has only been saved by private breeders. Man will not wait for the cooling of the world to consume everything in it, from teak trees to hummingbirds, and a century or two hence will find himself perplexed by a planet in which there is nothing except what he makes. He is a poor sort of creator."[11]

1926: OIL TO RUN OUT IN 1933

In 1926, the Federal Oil Conservation Board announced that the United States would run out of oil in seven years.

1948: "POPULATION OUTGROWS FOOD, SCIENTISTS WARN THE WORLD"

This was the front-page headline of the *New York Times* on September 15, 1948. The article said that the human race was growing fast, while natural resources were dwindling.

1948: WAR OVER NATURAL RESOURCES; SOIL EXHAUSTED

In 1948, two books appeared that seemed to confirm the gloomy *New York Times* prediction. Fairfield Osborn, the president of the New York Zoological Society, published *Our Plundered Planet*. Science commentator William Vogt's *Road to Survival* became a best-seller and book club favorite; he predicted that the depletion of soil and minerals would soon lead to smaller supplies and high prices, and possibly another world war, this one over natural resources.

1962: No Birds in the Spring

Rachel Carson, a biologist dying of cancer, warned the world that toxic chemicals were hastening the day when spring would come and no birds would sing.[12]

1967: Only Brutal Decisions Can Save Us

The entomologist Paul Ehrlich predicted imminent doom unless the government made "many brutal and heartless decisions": "the battle to feed all of humanity is over. In the 1970s the world will undergo famines—hundreds of millions of people are going to starve to death in spite of any crash programs embarked on now. At this late date nothing can prevent a substantial increase in the world death rate."[13]

1969: Fifty Years Left for Life on Earth

Barry Commoner, a biologist who ran for president in 1980, said that only a Socialist system could control technology and the greed that exploits it. In 1969, he predicted earth's life-support systems would be exhausted in fifty years.[14]

1972: Major Resources Gone by 1990s

The Limits of Growth declared that four forces, like the Horsemen of the Apocalypse, would end the world's joy ride. These forces were famine, the exhaustion of mineral resources, overcrowding, and pollution. The world's growing population would soon be fighting for food, making itself sicker with industrial pollution, and paying higher and higher prices for natural resources. The authors told readers that by 1981 the world's gold would all be mined. By 1985, its mercury would be gone. By 1990, no zinc; by 1992, no petroleum. By 1993, there would be no more copper for electric wiring or pennies, and no natural gas for home heating.[15]

1975: End of the World Might Be 1985

In 1968, George Wald, a Harvard biologist, won the Nobel Prize for his work in physiology and medicine. In 1975, he predicted that the world would end in twenty-five years, and later decided that the end might come as early as 1985.[16]

1978: "No Room in the Lifeboats"

This *New York Times Magazine* article warned that "the cost of natural resources is going up" and that we were entering an "Age of Scarcity."[17]

The editor of Britain's prestigious journal *Nature* summed up the main assumption of the doom promoters: "Their most common error is to suppose that the worst will always happen."[18] Of course, if the worst always happened, civilization would not have spread or survived. This kind of gloom is at the opposite pole from the conservation movement's faith in humanity.

By the end of the 1980s, almost no environmental writer had much good to say about the conservation movement. Their most common line was that conservationists achieved little and were simply part of the destructive system. At best, the conservationists had won some park designations and game regulations. At worst, they had aided and abetted the view of resources as human possessions by encouraging their efficient use.

The conservation and environmental movements are as different as two runners in a relay race. But the baton has been passed, and the new runner may be carrying it in a direction that would upset the founders of the conservation movement. The environmental movement is much less bound by science and economics, although it pays homage to both. In most ways, the conservation movement was a creature of science and economics. The sportsmen who founded it respected their quarry and enjoyed the wilderness, but they did not talk about animals as equal beings and they did not blame their religion, Western culture, or the political system for environmental destruction.

The Sierra Club's Paul Brooks was right when he said the movement had become revolutionary. But it was not the conservation movement. It was a new movement with different roots.

CHAPTER 3

An Opposition Movement Is Born

Finding the origins of the environmental movement is not so easy. In culture as in physics, every action has a reaction, but not necessarily equal and opposite. The idea that nature is a sacred book of moral wisdom was a reaction to the sudden power of science. The reaction starts in earnest in the middle of the eighteenth century.

For more than a hundred years in Europe and even in the boondocks of the American colonies scientists of the Enlightenment had been exploring the chemistry, physics, and biology of the earth. The sciences, rejuvenated in the fifteenth century when scholars picked up the pieces of Greek and Roman thought, were laying the final blocks in the foundation of the world's most profound transformation, the Industrial Revolution. Not long after Francis Bacon set down the principles of modern scientific investigation in the early seventeenth century, Isaac Newton, hiding in the country from a plague epidemic, conceived the ideas that enabled mathematics to explain the behavior of everything from a feather to the motion of planets. In 1789, a French lawyer-scientist, Antoine Laurent Lavoisier, completed the liberation of chemistry from the Greek and Roman theory of four basic elements (earth, air, fire, and water), and a brief-lived successor theory of a single essential matter called "phlogiston." In his *Traité élémentaire de Chimie*, Lavoisier declared that scientific experiment was the basis of all knowledge: "I have imposed upon myself the law of never advancing but from the known to the unknown, of deducing no consequence that is not immediately derived from experiment and observations."[1]

The public's near deification of Newton signaled that science had raised expectations of miracles. Also, the economic power to apply scientific discoveries had been growing. Looking back from the 1770s, the great theoretician of capitalism, Adam Smith, recognized the forces taking shape earlier in the century. Trade had become global. Riches were pouring into Europe. Between 1700 and 1750, England doubled the export of goods made from foreign raw materials. The power-multiplying inventions that symbolize the Industrial Revolution in high school textbooks took another century to appear—the steam engine, power looms, and railroads.

FEAR AND LOATHING IN FRANCE AND ENGLAND

No one sensed the transformation of science into industrial power with more dread and passion than Jean Jacques Rousseau, the French philosopher-poet who was wandering around Europe leaving his children on the steps of foundling asylums as he practiced being a "natural man." As we have learned in modern politics through the lives of the Kennedys and others, ignoble personal habits do not always diminish the influence of an otherwise bright and forceful personality and a fat pocketbook. Rousseau had the personality and brains, and he found the pocketbooks.

Rousseau dusted off the old debate about civilization and nature, human works and natural systems. The Enlightenment's passion for picking things apart and analyzing them seemed to him to be the last blow to nature. If *New Yorker* writer Bill McKibben had been on hand he might just as handily have used his title *The End of Nature* to describe the results of the Enlightenment. What troubled Rousseau and his peers was that if science could reduce all of nature to simple mechanisms and mathematical relationships, how could we know right and wrong? Was the suffering that one part of humanity inflicted on another simply nature's way?

Rousseau said that the answer came to him suddenly when he read the notice for an essay competition at the Academy of Dijon. The subject was the influence of science on morality. The prevailing wisdom of his time held that science and learning had inspired human progress and greater human happiness; no one knew that the Industrial Revolution was just around the corner, but science was ripe with big promises. Rousseau's 1750 *Discourse* said that this progress was an illusion, a trick. Anyone who really studied nature—not in little pieces, but as a whole—could see that primitive people who were innocent of science and its dissecting rationality had always led the happiest lives. The result of science and technology, Rousseau said, was greed. People grew accustomed to unnecessary luxury, which corrupted the simple virtues learned from nature. Yearning after material wealth led to some people having more than others. Science and technology had not saved humanity, but doomed it.

With a second *Discourse* in 1754, Rousseau won Dijon's honorable mention for his argument that when humankind had existed in animal innocence with nature, people had been "noble savages" (but, of course, not really savage at all). Men and women had lived in harmony with each other and with the world around them, and that meant they could do it again. And it could be done quickly, because according to Rousseau, every child is a noble savage. In the first sentence of his 1762 classic on the ideal education of children, *Émile*, Rousseau writes what today might be printed on an environmental poster: "God made all things good; man meddles with them and they become evil." This is the central theme of the environmental movement and also the substance of its permanent solution—restore the human psyche to innocence.

For Rousseau as for Vice President Al Gore, modern society is a dysfunctional family. It is worth recalling the vice president's words:

> Like the rules of a dysfunctional family, the unwritten rules that govern our relationship to the environment have been passed down from one generation to the next since the time of Descartes, Bacon, and the other pioneers of the scientific revolution some 375 years ago.[2]

By analyzing nature, science destroys its holistic beauty. By reducing nature to its working parts, science promotes the idea that we can improve and change the way nature works. Changing nature to our particular preferences encourages a preoccupation with consumerism. Material progress exemplifies the arrogance of the human family, which takes for itself what less able creatures cannot have. Greed takes over, each person wanting more and more goodies. Greed leads to economic injustice and the world divides into rich and poor.

What Rousseau feared and what environmentalists today still fear, and even despise, is not technology itself, but the human mind and civilization. Like the man who insists that all women should be as cleanly airbrushed as *Playboy* playmates, for the school of Rousseau, civilization's warts, occasional bad breath, and sweat make it unbearably crass and unclean. Anyone who doubts the durability of Rousseau's dislike for human society, science, and reason should pick up the thick summer 1993 catalog from the Omega

Institute. In one way or another, most of the courses in its ninety-seven pages promise to get Omega's affluent, educated customers away from the destructive world of reason and science and into the world of animals, nature, shamans, and Creation Spirituality, where the Dominican priest Matthew Fox "integrates the wisdom of nature, native cultures, and the Western mystical traditions with the scientific understanding of an emerging universe." Two popular Omegans from Australia, Michael and Treenie Roads, conduct "explorations into nature spirits," in which their students can "learn to commune intelligently and articulately with the souls of plants, animals, rocks, and rivers." The goal of Omega and most environmental theology is "to release our addiction to understanding."

Rousseau summed up the result of the human mind's meddling in another famous first sentence, this one in *The Social Contract:* "Man is born free, and is everywhere in chains." His words would come alive again a century later in Marx's manifesto: "Workers of the world unite, you have nothing to lose but your chains."[3] Rousseau's vision was revolutionary.

The French Revolution's rapid translation of noble sentiment into raging mobs, butchery, and dictatorship hardly dampened the idealism of Rousseau's followers, which had spread to England. (Idealists are like doting parents whose offspring never do wrong.) People who despise reason seldom use reason to analyze their failures. So it would be with communism in Russia and anti-Communist Nazism in Germany. Both movements, incidentally, relied heavily on the idea that peasant farmers and uneducated people were closer to nature.

How did Rousseau the eighteenth-century eccentric become part of mainstream America? Through England. There, more than in France, the Industrial Revolution was gaining momentum as people moved to the cities, a market system intervened between producer and consumer, and the first factories began to use specialized labor. In response to such trends, the spirit of Rousseau quickly became part of England's popular literature and scholarly learning in the age of Romanticism.

Mary Wollstonecraft seized upon Rousseau's faith in the natural person, but ridiculed his notion that women are naturally made to obey men; in 1792, *A Vindication of the Rights of Woman* launched the modern feminist movement. "That women at present are by ignorance rendered foolish or vicious," she wrote, "is I think not to be disputed." The remedy is to liberate the natural woman. "Would men but generously snap our chains, and be content with rational fellowship instead of slavish obedience, they would find us more observant daughters, more affectionate sisters, more faithful wives, more reasonable mothers—in a word, better citizens.[4] Unfortunately for her cause, Wollstonecraft died at age thirty-eight after the birth of her daughter.

That daughter would become Mary Shelley, wife of the poet and author of *Frankenstein,* in which she also picks up one of the great Romantic themes—how pride and greed inevitably create soulless scientific monsters. The subtitle of her book, *Or the Modern Prometheus,* refers to the Greek myth in which Prometheus steals fire from the gods and brings it to the human world. The story of science corrupting nature and bringing chaos into the world has had many incarnations, especially in movies. They include *Dr. Strangelove, King Kong, On the Beach, Planet of the Apes, Mad Max, The Andromeda Strain, War Games, Robocop,* and *Jurassic Park.*

The poets Percy Bysshe Shelley, John Keats, William Wordsworth, William Blake, and Samuel Taylor Coleridge came of age in the passions of Rousseau and the French Revolution. The Romantic mind is so much the mind of the modern environmental movement that tracing all the family lines would require a thick volume of intellectual genealogy. Most important is their shared dislike of science, technology, reason, business, and Western civilization. Like modern environmentalists, the Romantics contentedly enjoyed the blessings of the very things they condemned. Such logical contradictions did not bother them greatly. After all, logic was the enemy of real understanding.

When the painter Benjamin Haydon entertained Keats, Wordsworth, Charles Lamb, and other notable Romantics one evening in 1817, Wordsworth chided Haydon for painting the head of Isaac Newton. Newton, said Wordsworth, "believed nothing unless it was as clear as the three sides of a triangle." Wordsworth and Keats said that when Newton's simple prism experiments had explained the rainbow, he had destroyed its poetry. The poets carried the evening and the party offered a toast wishing confusion upon the science of mathematics.[5]

The Romantics were first among modern thinkers to legitimize the idea that profound inner truth far outweighs anything revealed by reason. Thus, the environmentalists' argument for preserving any natural resource is its intangible and unmeasurable value. Measuring benefits is cousin to materialism.

Before Wordsworth filled multiple volumes with galumphing verse in his middle age, he wrote lines that have become part of English and American culture. Many are still often quoted, particularly about modern materialism and the destruction of nature. Wordsworth was the first in the English-speaking world to substitute nature for God. Nature, he declared in "Tintern Abbey," should be our source of moral principles:

> In nature and the language of the sense,
> The anchor of my purest thoughts, the nurse,
> The guide, the guardian of my heart, and soul
> Of all my moral being.

In one of his sonnets, Wordsworth also gave the environmental movement the best distillation of its dislike for consumer society:

> The world is too much with us; late and soon,
> Getting and spending, we lay waste our powers:
> Little we see in Nature that is ours;
> We have given our hearts away, a sordid boon!
> This Sea that bares her bosom to the moon;
> The winds that will be howling at all hours,
> And are up-gathered now like sleeping flowers;
> For this, for every thing, we are out of tune.

Wordsworth and his friend Coleridge set out to change both the literary world and their society. In 1798, when the two published their anonymous *Lyrical Ballads* to challenge the literary elite with humble subjects, Wordsworth also included a famous challenge to traditional learning. He aimed "The Tables Turned" at his bookish friend, the essayist William Hazlitt:

> Books! 'tis a dull and endless strife:
> Come, hear the woodland linnet,
> How sweet his music! on my life,
> There's more of wisdom in it.
> . . . .
> One impulse from a vernal wood
> May teach you more of man,
> Of moral evil and of good,
> Than all the sages can.

Here in this English nutshell is the essence of Thoreau and his statement that has become the environmental movement's motto: "In Wildness is the preservation of the world." Here is the faith that creates the passion for saving urban greenways, tall-grass prairies, and redwood forests. Through the perspective of this faith, anyone who cuts down a big tree or drains a marsh is either stupid or evil.

After World War II, the Romantic vision gained more and more converts: beatniks, then hippies, and soon a whole movement began to see nature as perfect, humanity as corrupt. We began to believe that we were not to blame for our failures and illnesses. The problem was that we had abandoned nature and its laws. Nature became the alternative to civilization and civilized behavior. Just as the modern environmental movement was in its second trimester of gestation, one of the most influential youth rebellion movies of the 1960s took its title from a Wordsworth line and wove his poem into its plot. In *Splendor in the Grass*, a young girl struggling to free herself from her parents' traditional American values clings to Wordsworth's "Ode: Intimations of Immortality." The poem begins in a despairing tone familiar in environmental writing: "But yet I know, where'er I go, / That there hath past away a glory from the earth." It also includes the idea that in being born and civilized we have lost the Golden Age:

> Our birth is but a sleep and a forgetting:
> . . . .
> Heaven lies about us in our infancy!
> Shades of the prison-house begin to close
>   Upon the growing Boy.

The poem ends in the faith that all the sorrows of the civilized world, all the tears, can be conquered by the power of nature if we let it in: "To me the meanest flower that blows can give / Thoughts that do often lie too deep for tears."

Anyone who doubts that the Romantics shaped popular opinion on science, business, and industry for all time should consider the difference between the reality of nineteenth-century England and the England that has come down to us from Romantic writers and their Victorian followers. Every American student learns that the Industrial Revolution caused immense social upheaval and suffering. For environmentalists, the Industrial Revolution is the beginning of humankind's destructive behavior, or at least willfully destructive behavior. Its smokestacks, which Blake called "dark Satanic Mills," were the forward gun batteries in the assault on nature.

Like environmentalists and most liberals today, the Romantic and Victorian writers condemned the present, romanticized the past, and predicted disaster for the future. Who more than Charles Dickens's Tiny Tim, David Copperfield, and Nicholas Nickleby symbolize the human side of the Industrial Revolution and its swelling urban masses? It is no accident that Karl Marx and Friedrich Engels came to England at this same time to write the books that for more than one hundred years would provide the basic arguments for almost every group that blamed capitalism for the ills of society and nature. They would also eventually lead to the most tragic failure of social engineering and the greatest rape of nature that the world has ever known.

In the nineteenth and twentieth centuries, the attention focused on human suffering and environmental damage has saved both lives and resources, but it has been kind to only a few lucky individuals and groups. A natural outgrowth of the Romantic preference for a personal relationship to nature, and to most of the world, may be to emphasize individual cases rather than broad trends and concepts. Romantics, and their descendants in the environmental movement, mastered the use of the anecdote of the exceptional case. The plight of a single person, family, or group comes to stand for the failure of the whole society. To bend a familiar saying, environmentalists tend to focus on trees, while scientists and economists focus on forests.

The deadliest seed the Romantic movement planted was the idea that real heroes fight society, like the revolutionaries of France and America, or else they spurn society like life's outcasts—tramps, hobos, misfits, and eccentrics. Goethe's melancholy Werther lives on the brink of suicide, blaming society for his alienation. Wordsworth's beggars and crazy people are always wiser and more virtuous than society. Dickens's children and poor are generally more virtuous than their parents or the rich. When a Romantic hero is defeated, the fault is never his own. Unlike earlier heroes of tragedy—Lear, Hamlet, or Oedipus—modern heroes and environmental heroes are victims. Even in defeat they do not admit that they might have controlled their fate by wiser action.

As the American environmental movement gathered strength in the 1980s along with the economy, its voice grew louder, saying things were getting worse. Human vices were ruining both nature and society. The rich were getting richer, the poor, poorer. This is a story learned from the English Victorians, who focused intently on sections of society that were in fact getting poorer, at least for a while. Profound economic and social change always leaves some people behind. Contrary to President Reagan's happy assumption that the rising tide lifts all boats, boats with short anchor lines disappear beneath the waves. Yet such suffering is not irreversible, and it is not a reason to assault a system that benefits the great majority.

THE FIRST GREAT DOOMSDAY THEORY

The ghost of the gloomiest of all the Romantics is still channeled through the mouths of environmental media stars such as Paul Ehrlich and Carl Sagan. The Reverend Thomas Malthus, born in 1766, grew up in an upper-middle-class family. He was educated by his father and by politicized tutors, one of whom was jailed for inviting the French revolutionaries to bring their murder and mayhem to England. Malthus remained an academic his entire life. He was passionately concerned with the plight of the poor. He felt the wrongs of the present painfully, but he could only theorize about the future, since he understood little of the forces changing his world. He offered a patently unrealistic solution—"we should facilitate instead of foolishly and vainly endeavoring to impede, the operations of nature in producing mortality."[6] In other words, let the poor and sick die off.

Malthus formulated the first great doomsday theory, one still being preached from environmental pulpits. According to Malthus, "the power of population is indefinitely greater than the power in the earth to produce subsistence for man."[7] He warned that population growth was already outrunning resources. Don't be misled by the optimism of scientists, Malthus declared. Their revelations are brilliant, but useless in avoiding catastrophe. The earth's carrying capacity was being strained, he said, and the world must begin shedding its weight of people.

Malthus could see little but the world nearby in time and space. The discussions with his father that led to his anonymously published "Essay on the Principle of Population" took place when poor harvests and war raised British food prices to record levels. His grouchy view of the human mind prevented him from realizing that human ingenuity in the Western world was already multiplying productivity more rapidly than the population could consume the results. He could see only the poor, who stood for the failure not only of science and industry but of society's potential for charity as well. Sounding like a befuddled member of President Lyndon Johnson's Great Society program in the 1960s, Malthus wrote, "notwithstanding the immense sum which is annually collected for the poor in this country, there is still so much distress among them." And just in case increased demand did lead to increased productivity, "the spur that these fancied riches would give to population would more than counter-balance it [productivity]."[8] Exactly the opposite happened and continues to happen; the one sure form of birth control is increased productivity and affluence.

If science and industry had been stopped in the mid-nineteenth century, the world population would have outstripped nature's support systems. Instead, we modified those systems. For more than 100,000 years, humankind had been using technology to increase the earth's carrying capacity faster than it has increased its own numbers. Early hunters and gatherers, who used only clubs and sticks, probably needed several hundred acres per person to survive. The stone-tipped spear and arrow reduced the acreage a little, and agriculture probably reduced it to an acre or two per person. Technology has played a much greater role in increasing carrying capacity than in decreasing it.

Rousseau had it backwards. Human beings are born in chains, but we have freed ourselves from both nature's limits and the limits of our own ignorance. Of course, nothing about the human mind guarantees that we will continue to solve every problem or that we won't make colossal mistakes. As we become more numerous and powerful, our freedom to make mistakes becomes more frightening. But it is the fear of this freedom that is likely to put us in chains again.

WHAT REALLY HAPPENED IN
NINETEENTH-CENTURY ENGLAND?

If the Romantics and Victorians had not succeeded in eclipsing a rational assessment of their own time, Marx and Engels might have been little more than brilliant but dreary philosophers. Europe might have been happier, and our own social movements might have learned how to build on strengths rather than building expensive and extravagant defenses against weaknesses.

A society that allows five- and ten-year-old children to work at mechanical looms, men to work stooped over in four-foot-high tunnels, hacking all day at rock walls, and women to crawl up and down mine shafts laden with coal needs powerful reform. The Wordsworths, Blakes, and Dickenses hurried reform along by humanizing and even romanticizing the poor. Their work was visible and readable. Yet much more powerful, if less colorful, forces were also changing society for the better. Just ask where the world would be without the following inventions: the McCormick reaper, Fulton's steamboat, the railroad, the telephone. In the nineteenth century, "natural" folk medicine such as applying leeches, purging, and bleeding was replaced by bed rest and human attention. Medical progress resulted in dramatic improvements in infant mortality. In the 1700s, Adam Smith reported that many women in the Scottish highlands were able to rear only two children out of twenty. Women were often pregnant from puberty until death. These harsh circumstances came not from industry, but from the inability to control the natural environment. This was a time when 80 percent of England's population lived in the "healthy" countryside.

By the time Wordsworth and Coleridge published *Lyrical Ballads,* claiming dignity for the poor and the uneducated, the national economy was already opening the doors for a dignity beyond poetic blessing. By 1800, the growth rate of the national product decisively outstripped population growth for the first time in a thousand years or more. By the 1830s, an economic growth rate of 2.9 percent promised to double in twenty-five years. Incomes were rising about 1.5 percent a year, or doubling in fifty years. This may seem slow, but it is faster than the general rise in living standards during the preceding two thousand years. And with many ups and downs, the living standard would continue to improve. Try to find a single environmental book or magazine that publicizes this improvement in human life. These publications are full of graphs projecting species extinctions, the population explosion, the exhaustion of fossil fuels, and global warming. Almost all these predictions are computerized guesses that extend past trends into the future as if nothing can change for the better. Especially not human life.

When my gloomier friends in the environmental movement tell me how much worse life has become, I always ask them how they measure this. If they say it is subjective, I reply that it must be based on the observation of something we have all seen, and what might that be? Set aside the special and usually short-lived situations such as the sudden unemployment of horse handlers when the automobile arrived. Draw a smooth line through the ups and downs caused by war, disease, weather, and business cycles. By anything we can see or measure, life in the Western world began to get dramatically better around 1800. For example, a woman born in 1800 could expect to live no longer than forty years. A child born after 1945 could expect to live almost twice as long. Food per person in the world has risen continuously, even if it is not well distributed. The prices of essential metals and fuels have all fallen. The Industrial Revolution was a mixed blessing, but name a blessing that isn't mixed. In the span of history, the revolution's worst abuses were short-lived outside the Communist world, and even the staunchest environmentalists enjoy its blessings, while advertising its general failure.

Like the genes that determine our biological character, this European story I have told determined the character of the American environmental movement. The movement is thoroughly Romantic in its rejection of science and reason, and thoroughly Victorian in its humorless moralism, dark assessment of human nature, and belief in the inevitability of suffering.

America could not become fertile ground for this vision of the world until it bred a class of intellectuals as affluent and privileged as English ladies and gentlemen, and as generally removed from the world's daily work. However, when conditions were right, these essentially un-American notions quickly made their way across the Atlantic and were welcomed and preserved by a literary elite.

CHAPTER 4

The Roots of Environmental Thinking in America

People who profess to love nature above all things are usually selling something. It is no accident that the only people to rival the Romanticism of environmentalists on the subject of nature's goodness are the commercial hucksters of the sixteenth and seventeenth centuries, who tried to lure new settlers to America to farm, mine, trap, and cut timber. Their advertisements for colonists extolled a land of friendly natives and nature so rich that it seemed like a banquet table and give-away jewelry store. An easy fortune, good health, and a long life awaited anyone smart enough to escape the remnants of medieval Europe. Columbus and other explorers who did not have to cope with living in the wilderness or who wanted to finance their next voyage or recruit settlers also often promoted the Garden of Eden idea. When Columbus wrote of the crystalline waters of the southern Caribbean, he speculated that explorers would find their source to be the original Eden.

The bubble of any such fantasy burst after a day or two in the wilderness, however. The English settlers who arrived on the *Mayflower* in November 1620 stepped ashore into what their leader William Bradford called a "hideous and desolate wilderness, full of wild beasts and wild men."[1] This is a narrow view of a rich continent, but danger and disaster did surround the first European settlers. Survival on every part of the American frontier meant the immediate conquest of nature. Becoming prosperous, or even surviving for a normal lifetime, meant constant struggle against weather, wild animals, weeds, and isolation.

The history professor Leo Marx says that the practical experience of taming a wilderness lies at one extreme of the American imagination, and at the other extreme lies the continuing romance of nature as a garden, the pastoral ideal.[2] Marx reminds us that these two views—the pleasant garden and the dangerous wilderness—had real consequences: "In other words, each image embodies a quite distinct notion of America's destiny—a distinct social ideal."[3] The image of the garden, whether wild or cultivated, celebrates nature's virtues and its ability to provide for human comforts. The image of the dangerous and hostile wilderness inspires efforts to exercise our powers, "to postpone immediate pleasures, and [to] rehearse the perils and purposes of the community":

> Life in a garden is relaxed, quiet, and sweet . . . but survival in a howl-
> ing desert demands action, the unceasing manipulation and mastery of
> the forces of nature, including, of course, human nature. Colonies
> established in the desert require aggressive, intellectual, controlled,
> and well-disciplined people. [Note that Marx does not say moral, ethi-
> cal, or virtuous people.][4]

American history has been dominated by the challenge to master and manage nature. The environmental movement and its historians like to picture the loggers, farmers, miners, and developers from colonial days to the nineteenth century as people who thought America's resources were endless. Many of these early Americans, however, cared deeply about both the utility and beauty of the country's natural resources. From this tradition arose the conservation movement.

The first botanical gardens in America were established a few years before the American Revolution. Their inspiration was scientific, aesthetic, and practical. Benjamin Franklin designed his Franklin stove in part to assure less waste of wood. He helped the botanists John Bartram and Humphrey Marshall find plants to increase the productivity of farms and forests. And it was Franklin who recommended lining all the streets of Philadelphia with trees. Dr. Nicholas Collin, a pastor in Philadelphia, addressed the powerful American Philosophical Society in 1789, recommending that the colonies adopt programs to preserve forests for fuel and other domestic uses. In laying out the grounds of Monticello, Thomas Jefferson decided that harmonizing his mountainside home with nature required that he limit himself to indigenous plants. In 1832, the painter and buffalo hunter George Catlin proposed a great national park in the West where might in the future be seen . . . preserved in their pristine beauty and wildness, in a magnificent park, where the world could see for ages to come, the native Indian in his classic attire . . . amid the fleeting herds of elks and buffaloes. What a beautiful and thrilling specimen for America to preserve and hold up to the view of her refined citizens and the world, in future ages.[5]

In 1858, Thoreau noted the European tradition of game preserves and deer parks and wrote in the *Atlantic Monthly:* "Why should we not . . . have our national preserves . . . in which the bear and panther, and some even of the hunter race, may still exist and not be 'civilized off the face of the earth.'"[6]

What really winnowed the conservationists from the Victorians holding the smoky tinder of the environmental movement were two books published within four years of each other. In 1859 in England, Charles Darwin, after a life of observing nature around the world and years of meditating on its variety, published *On the Origin of Species.* In the public mind, Darwin's book was quickly reduced to one central idea—that humankind is not a special creation of God as humans are and have been in almost every religious story from the most primitive to the most literate. However nature originated—in the mind of God or a big bang—Darwin's evidence made human beings just one of many beings, a part of nature. From the time of Darwin humanity had to recognize that with or without the knowledge and participation of God, we were as much subject to nature and its forces as a fish or a redwood tree. In fact, the rise of our civilization and culture seemed irrevocably tied to changes in the earth itself.

Five years after Darwin's *Origin of Species* delivered its humbling message, George Perkins Marsh, an American born in the near wilderness of frontier Vermont, described the historical consequences of thinking ourselves apart from nature. As the U.S. ambassador to Turkey, Marsh combined his diplomatic duties with long explorations of the Middle East. His treatise on camels would lead the U.S. Army to experiment with camels in its western campaigns. Almost everywhere he went, Marsh, like today's tourists, found the ruins of the biblical and classical world fascinating. For this man from the wilderness, they came to be the basis of a profound realization.

While serving as President Lincoln's ambassador to Italy during the Civil War, assuring a continuous supply of war materials to the Union, Marsh had his epiphany. These old civilizations, he realized, had contributed to their downfall by devastating their own environment. The deserts of North Africa and the Middle East, the barren hills of Greece, and the vanished forests of biblical lands were all evidence that human societies, thinking themselves special, were capable of destroying their own life-support systems. In 1864, Marsh published the most important conservation book of the nineteenth century, *Man and Nature*. From its long catalog of evidence emerged a single great question: how can America avoid the fate of Egypt, Mesopotamia, Greece, and Rome?

THE DARK VIEW CROSSES THE OCEAN

Neither Marsh nor Darwin were Romantic pessimists. They were scientists. They believed in Western values and culture. Marsh did not suggest stopping the progress of industry in America or aiming for a simpler life. He set the tone of American conservation for the next century—scientific evaluation and prudence.

The dark view of human nature that was flourishing in Germany, England, and France and that would eventually surface in the modern environmental movement found shelter among American academics and intellectuals who were outside the mainstream. Most were people who, by the fortune of their birth or social connections, never had to compete with nature for a living on a farm or in a forest, and never had to compete with other humans for a living in the business world. They liked to declare how uniquely American they were, but except for their subject matter, they thought like the English Romantics and Victorians.

Ralph Waldo Emerson positioned himself as the commander of the war for American intellectual independence from Europe. Yet Emerson's writing is sometimes a mirror image of Marx's views on how industrialization and mechanization replace the happy rural farmhand with the alienated worker. And with somewhat less elegance than the original, he paraphrases Wordsworth's sonnet "The World Is Too Much with Us": "Things are in the saddle / and ride mankind."

Emerson remained more in love with the idea of nature than with nature itself. The historian John Reiger calls Emerson and his contemporaries "voyeurs."[7] In 1871, Emerson disillusioned the ebullient wilderness trekker and Sierra Club founder John Muir when he refused to spend a month in the Sierra wilderness, or even a single night among the giant sequoias. The notes Muir wrote in the margins of Emerson's works show his despair about Emerson's ignorance of animals, plants, and rocks as they look and function in real life.

Emerson was not interested in science because science limited itself to information collected and verified by the five senses. A former Unitarian preacher, Emerson was more interested in matters of the spirit. Reacting against scientific rationalism, he developed a new faith, Transcendentalism. The only requirement was to believe "in an order of truth that transcends the sphere of external senses."[8] Transcendentalism's famous followers included New England's best writers: Margaret Fuller, Nathaniel Hawthorne, and Bronson Alcott. Their faith that the soul could communicate directly with nature found validation in new translations of the oriental mystics. Environmentalists today continue to believe that Eastern mysticism offers a superior understanding of nature to Western science; however, they seldom check the results of these belief systems against the reality of how Eastern cultures have treated nature.

Emerson named and framed the new vision, but the messiah who showed and told America how to live it was a young, eccentric friend of Emerson whose words would become environmental gospel. Henry David Thoreau is remembered most for two actions—retreating to a shack on Walden Pond and spending a night in jail for an act of civil disobedience. These are hardly distinguished actions, but the way he wrote about them was brilliant. His finely wrought prose sparkles with wit. It inspires a reader to try to know nature through every sense. It was great writing, fine observation, and sloppy thinking. His prose is so good English professors teach Thoreau with their critical faculties in suspended animation. Many have swallowed his silliness along with his sensibilities.

It is worth looking closely at Thoreau because he remains the messiah of the environmental movement. The son of a storekeeper turned pencil maker, Thoreau left Harvard and did what many affluent American students did in the 1960s and 1970s. He drifted from job to job. He became a nature writer after brief attempts at teaching. He went backpacking and finally went "back to the land" at Emerson's Walden Pond, dropping out of society to find deeper meaning in a natural setting.

Thoreau observed nature closely, but not for the understanding that a scientist seeks. Thoreau had already turned away from the Western mind. He had read the Eastern classics distributed today at airports and on city streets by Hari Krishnas in orange robes: the Vedas, *Mahābhārata,* and *Rāmāyana.* These texts direct the mind and attention inward. Contact

with nature is sought not to understand nature, but to create an inner subjective sense of harmony with nature, and ultimately an escape from the
everyday world.

Under these conditions, nature becomes a friend, a close relative,
Mother Nature, or Gaia. Thoreau describes a gentle rain in which there was

> such sweet and beneficent society in Nature, in the very pattering of
> the drops, and in every sound and sight around my house, an infinite
> and unaccounted friendliness all at once like an atmosphere sustaining
> me, as made the fancied advantages of human neighborhood insignifi-
> cant, and I have never thought of them since. . . . I was so distinctly
> made aware of the presence of something kindred to me, even in
> scenes which we are accustomed to call wild and dreary.[9]

With a railroad running by one end of the pond, a road a few hundred
yards away, and friends at the ready in Concord, Thoreau could afford the
luxury of being pleased with everything nature did. If rain rotted the seeds
and potatoes in his garden, he said, it would be good for the hillside grass
and thus for him.

When he went hiking, his closest companion says he generally stayed
on the roads and paths, and he was sure to have with him not flour, but
bread, pork, guidebooks, salt, sugar, tea from India, and rubber boots as
well. This is hardly meeting nature on its own terms, staking one's life on
the good will of the wilderness, or sharing the American frontier experience. The books that Thoreau wrote about his experiences did not appeal
to Americans preoccupied with nature's real challenges or struggling to
create cities and jobs.

Seizing skillfully on the problems of his time, Thoreau used the
Romantic tactic of turning the specific into the general, his personal experiences into universal truths. The value of literature, we are told, is that it
speaks to people of all times. This is also its danger. The final achievement
of Thoreau's dazzling prose has been to blind many people to his society's
most enduring accomplishments. Thus, ideology triumphed over sense and
science. This is easy to demonstrate by quoting a few beloved passages and
extracting the principle they embrace:

The child is smarter than the adult.

"I have always been regretting that I was not as wise as the day I was
born. The intellect is a cleaver; it discerns and rifts its way into the secret of
things."[10]

We learn more looking inward than outward, indulging in our
own feelings than exploring the world on its own terms.

"It is easier to sail many thousand miles through cold and storm and cannibals, in a government ship, with five hundred men and boys to assist one, than it is to explore the private sea, the Atlantic and Pacific Ocean of one's being alone."[11]

The poor are really happy and wise.

"Love your life, poor as it is. You may perhaps have some pleasant, thrilling, glorious hours even in a poorhouse. . . . The town's poor seem to me often to live the most independent lives of any. Maybe they are simply great enough to receive without misgiving."[12]

Nature is wise and deliberate; humankind is confused, frivolous,
and misled.

"Let us spend one day as deliberately as Nature, and not be thrown off the track by every nutshell and mosquito's wing that falls on the rails."[13]

Technology blinds us.

"Men think that it is essential that the Nation have commerce, and export ice, and talk through a telegraph, and ride thirty miles an hour . . . but whether we should live like baboons or like men, is a little uncertain." He proceeds to lampoon the nation's fascination with railroads, suggesting that people believe the railroad will eventually carry them to heaven: "But if we stay at home and mind our business, who will want railroads? We do not ride on the railroad; it rides upon us."[14]

Industry exploits innocent people.

Thoreau not only despises the railroad, but condemns industry for sacrificing human lives to its ambitions. "Did you ever think what those sleepers are that underlie the railroad? Each one is a man, an Irishman, or a Yankee man. The rails are laid on them, and they are covered with sand,

and the cars run smoothly over them. . . . If some have the pleasure of riding on a rail, others have the misfortune to be ridden upon."[15]

Consumer goods corrupt rural life.

When Thoreau hears the train go by Walden Pond, he thinks of all the goods it is bringing to the city: "All the Indian huckleberry hills are stripped, all the cranberry meadows are raked into the city. Up comes the cotton, down goes the woven cloth; up comes the silk, down goes the woolen; up come the books, but down goes the wit that writes them."[16]

Private property created by greed corrupts and enslaves its owners.

Thoreau boasts of all the farms he has turned down. For him, rejecting private property is rejecting slavery. He swears to be buried before he will own a farm. "I would say to my fellows, once for all, as long as possible live free and uncommitted. It makes but little difference whether you are committed to a farm or the county jail."[17]

Nature produces better medicine than science.

The best medicines, Thoreau writes, are "nature's universal, vegetable, botanic medicines." He would take fresh morning air over pills any day. He worships the wild plants that have "the power of restoring gods and men to the vigor of youth."[18] Thoreau, like the average person of his time, died at age forty-four, of tuberculosis, nature's victim.

PRACTICAL ROMANTICISM = SOCIALISM

In France, England, Germany, and America, Romantics looked at the chaos, filth, and misery of the industrial boom towns and imagined alternatives. *Imagined* may be the wrong word, since all the alternatives were variations on a single theme—rural communal harmony.

In England, the poets Samuel Taylor Coleridge and Robert Southey planned an American socialist utopia. Southey wrote his brother that he and Coleridge "preached Pantisocracy and Aspherism everywhere. These,

Tom, are two new words, the first signifying the equal government of all, and the other the generalization of individual property."[19] Like America's "back-to-the-landers" of the 1960s and 1970s, these affluent university boys imagined a rural commune where twelve men and twelve women would work only two to three hours a day. The rest of the time they would commune with nature on the banks of the Susquehanna River and improve their minds. The plan hit a fatal obstacle when Coleridge discovered that Southey did not intend to put his personal wealth in the communal pot and that nothing would be held in common but five or six acres of farmland.

A much more practical Welsh businessman, mill owner, and social reformer actually did commit his fortune to buying American land, where communal life surrounded by nature would show how the right environment could "remove all causes for contest between individuals" and cleanse human society of evil.[20] The fame that Robert Owen had won by his humane treatment of his company's mill workers quickly drew one thousand utopians (including a few rascals) to his newly purchased community of 20,000 acres of farmland in Indiana. New Harmony began in 1826—and had fallen apart by 1830.

Emerson and Thoreau never joined one of the new communes, but they occasionally appeared at Brook Farm on the Charles River near Boston. Nathaniel Hawthorne joined only to find that after a day of farm labor he could not write. The true Transcendentalist commune was Bronson Alcott's 160-acre Fruitlands. Alcott espoused almost every principle held by an environmental group. In the few months that the commune existed, Fruitlanders practiced vegetarianism and abstained from milk produced by imprisoned cattle. They despised commerce. They would not stay up late burning animal oil in their lamps. Like good "bioregionalists," they dined on nuts, fruits, and herbs native to the area.

These nineteenth-century keepers of the Romantic flame had little impact on the America of their time. Their ideas were not taken up on a larger scale until the early twentieth century, when the founders of the environmental movement came on the scene. Two events gave them the opportunity to seize the initiative from the conservation movement. The first was World War I and the second, the Great Depression of the 1930s.

During the war, for the first time technology demonstrated its power for massive destruction of humanity and nature. Photography and cinema brought home the war-stripped forests of the European front as well as the images of thousands of men cut down in a single battle by cannons and automatic weapons. In the Great Depression, capitalism's general progress toward greater prosperity seemed to have failed. A movement called Technocracy blamed unemployment on modern labor-saving technology. The movement cited case after case of workers losing high-paying jobs and surviving on lower-paying, less skilled jobs. The Depression made earlier times and simpler life seem ideal.

The single most powerful work of history during this time appeared shortly after the war, when a wealthy German philosopher, Oswald Spengler, wrote *The Decline of the West*. His notion that Western civilization was choking on its success helped explain the Depression. The West's decay, Spengler wrote, was symbolized by its large cities. This idea proved popular with the men who would lay foundation stones for the modern environmental movement.

In 1933, Robert Marshall, heir to millions, hiker, forester, and founder of the Wilderness Society, advocated nationalizing much of the country's private forest lands so that citizens could escape modern life: "As society becomes more and more mechanized, it will be more and more difficult for many people to stand the nervous strain, the high pressure, and the drabness of their lives. To escape these abominations, constantly growing numbers will seek the primitive for the finest features in life." The nation, Marshall wrote, "cannot afford to delay their [the forests'] nationalization until the form of government changes."[21]

Socialism was Marshall's answer. He joined Norman Thomas's Socialist party while his friend and Wilderness Society cofounder, Benton MacKaye, joined Eugene Debs's Socialists. MacKaye, like the English Romantics and Rousseau, saw the city as the great evil. He spoke of the "wilderness of civilization" and the "slum of commerce."[22] Social engineering was his solution, and it shows the shortsightedness of prescribing fixed answers in a changing world. MacKaye, for instance, was a great fan of accommodating the car on freeways, now the nightmare of environmentalism.

When the environmental movement got its first presidential candidate, the biologist Barry Commoner, in 1980, it is significant that he was both a Socialist and a believer that science is dehumanizing. Like modern critics of biotechnology, Commoner was repulsed by Francis Crick and James Watson's work on genetic codes. Like environmentalists who build a future on the foundations of a fanciful past, Commoner lamented the shift from biodegradable wood and cotton to synthetics. Hell-bent on undermining modern science and commerce, he could not admit that relying on wood, cotton, and other natural substances had devastated large areas of the environment, while synthetics brought more and cheaper goods to the world's poorest people.

According to Commoner, we had "broken out of the circle of life." His 1971 best-seller, *The Closing Circle*, estimated that the world had twenty to fifty years before it was irreversibly ruined. He campaigned for the presidency as a Populist, but his emphasis on the wisdom of central government to mandate change demonstrated once again the appeal of socialism to environmentalists.

When Dennis Hayes took leave from his job as a congressional staffer to organize the first Earth Day in 1970, he did not hide his economic philosophy: "We demand a lower productivity and a wider distribution."[23] David Brower, once president of the Sierra Club and Friends of the Earth, helped organize an environmental conference in Nicaragua near the end of the Socialist Sandinista era. Despite the continued destruction of that country's environment and the government's use of environmental laws to turn a nature preserve into a military airfield, Brower declared that Nicaragua was a model for the capitalist world.

With the failure of the great Socialist experiments in Eastern Europe, Africa, and Latin America, and with other countries privatizing as fast as possible, the word *socialism* has disappeared, but the rhetoric and world view remain. Most people in the environmental movement still see free-market democracy through the eyes of Rousseau, Thoreau, Marshall, and MacKaye. *Buzzworm,* an environmental journal, worried in June 1993 that the Clinton-Gore program had evoked a dangerous reaction; in the usual good guys/bad guys division of the world, the editor Joseph Daniel warned, "Meanwhile, the opposition—those black knights of rampant greed—are regrouping."[24]

Today's most fashionable idea among environmentalists and foreign aid agencies is "sustainable development." Impoverished countries wanting to enjoy the comforts of civilized technology can avoid the natural plunder of the industrialized West, they say, by practicing sustainable development. It is supposed to be part of a new model of the relationship between humanity and nature. The central idea is to use nature without injuring its capacity to support future generations at least as well as it supports our own. Its champions offer a variety of recipes. Common to all the recipes is one ingredient—government distribution of resources and centrally planned limits on technology and industry.

But aren't such limits necessary? Yes, and they may be imposed in two ways—by a select few who run the government or by the marketplace. A large body of free-market environmental solutions has begun to vie with the authoritarians for attention, and the contrast shows how sustainable development leans toward old Socialist solutions. If disposing of plastic and newspapers in landfills costs, say, one dollar per garbage can, free-market environmentalists say that citizens should be able to choose whether to recycle, reduce their use of such products, or pay the dollar per can. Most environmental groups propose only government-mandated recycling. Sustainable development disciplines from the top down. It depends on government officials being wise and setting aside any self-interest. It seeks population control. Since it does not expand resources, it must appease those at the bottom of the economic ladder by taking from those at the top. The proponents of sustainable development insist that it is compatible with nature. On the contrary, the first thing it does is ignore *human* nature in much the same way that the Soviet empire tried to ignore human nature for seventy years.

While environmentalists believe the natural person is good, in a society divorced from nature, humankind is inevitably untrustworthy and unstable, given to greed and violence. In the first surge of environmentalism in the early 1970s, communes sprang up all over rural America. Their members vowed to live simply and share their wealth. Some survived for a decade, most no more than a year. Although socialism has been largely discredited since the 1980s, a remnant lives on in the environmental movement's preference for centrally mandated and policed environmental regulations. It also survives in the belief that the consumption of resources by rich people and rich countries unfairly deprives the poor of life's necessities. Environmentalists still believe that the solution to human and environmental problems is to redistribute the world's wealth by government decree, as evidenced by the 1992 Earth Summit in Rio de Janeiro and by their backing for President Clinton's "tax the rich" platform.

In the simplest terms, the environmental movement believes that greed destroys nature by consuming its resources without thinking about the consequences for tomorrow. Thus, environmentalists constantly cite figures that show the United States has 6 percent of the world's population yet consumes 50 percent of its resources. ("Consumes" is never defined, nor the source ever noted. No one bothers to mention how much the United States pays to people who provide the resources or how, once "consumed," the global market distributes them around the world.)

Although environmentalists know little of history in general and even less of economic history, they have a vague outline that brings them to their present dislike for Western capitalism and free-market democracy. This sketchy, distorted view of history adds up to something such as Audio Literature's 1992 cassette by Wendell Berry, *What Are People For?* Perhaps its silliness is obvious, but it invites commentary, since Berry, on his Kentucky farm, is the movement's living Thoreau. Here is history by a lion of environmental literature, with my notes:

One cannot maintain one's competitive edge if one helps other people.

Berry apparently does not know that corporate charitable contributions *to whom? gout* during the 1980s grew faster than during the previous twenty-five years *officials?* and became a larger part of corporate profits. *how did char. contributions make profits -tax write off? Enron.*

The ambition to feed the world or to feed the hungry, rising as it does out of the death struggle of farmer with farmer, proposes not the filling of stomachs but the engorgement of the bottom line.

Perhaps Berry could explain this one in Somalia.

All of rural America sits as if condemned in the shadow of the free market.

A few million Russians are ready to come and save it, not to mention savor it.

Competition is destructive both to nature and human nature because it is untrue to both.

The famous evolution or "monkey" trial took place in neighboring Tennessee in 1925, and it seems clear which side Berry is on.

The defenders of competition have never known what to do with the losers.

Tell that to Japan, Germany, and Italy, and to the new businesspeople of the former Communist world enjoying their World Bank loans.

Every transaction is meant to involve a winner and a loser.

Did you feel that way when you bought your last car or bag of organic vegetables? Maybe Berry felt that way when he got tenure.

It [capitalist society] has done by insidious tendency what the communist revolutions have done by fiat. It has dispossessed the people and usurped the power and integrity of community life.

Which must be why Americans turned out in record numbers to vote the year this was written and why millions of immigrants from Russia and Vietnam try to enter the United States every year.

Berry's formula of history appears over and over in literature from the environmental movement. Shortly after the first Earth Day, a group of little-known scientists calling themselves the Club of Rome published a sensational Malthusian declaration, *The Limits to Growth*. The club based its cry of alarm on simplistic computer modeling and assumptions that human beings could not adapt to changing economic conditions. The book was quickly followed by an attempt to give it scientific standing, when the editors of *The Ecologist* came out with *Blue-Print for Survival*. Its introduction repeats the standard Malthusian formula: "*Indefinite* growth of whatever type cannot be sustained by *finite* resources. This is the nub of the environmental predicament."[25]

I cite here just two of their predictions, followed by the reality in each case:

- Demand for gasoline will exceed supply by the end of the century. (In 1993 gasoline continued to sell at inflation-adjusted prices close to 1970 levels.)
- If present rates of expansion continue, no currently marginal land will be unfarmed by 1985. (In the 1980s and early 1990s the European Community countries and America as well as Vietnam and India were producing surpluses of many crops, and America was taking land out of production.)

The members of the club and the editors of *The Ecologist* seem unaware of the human animal's most valuable talent. While other animals adapt to environmental change by chance, the human animal has learned to adapt quickly and deliberately by reason. Ecologists who assume the absence of adaptation are not ecologists in any modern sense of the word. The Club of Rome's report and its endorsement by *The Ecologist* are not works of science, but of fear and politics.

The environmental movement's solution to the imaginary crisis in resource supplies and to the real but exaggerated deterioration of some ecosystems has always been to regulate the use of both private and public resources. The movement has actually succeeded in taking away the rights to certain activities such as cutting timber, plowing land, building, and draining wetlands. Today, the regulations routinely supported by the environmental movement amount to public ownership of private property. When Ronald Reagan became president, the greatest environmental controversy was over his proposals to return rights to private property owners by reducing regulations and to expand the private control of economic and natural resources by privatizing government property and business. I will discuss both proposals later in detail; I mention them here to point out that, in essence, environmental politics are a debate shaped by two different versions of history.

WHERE ENVIRONMENTAL
HISTORY LEADS

The environmental movement reinforces its version of history by asking other people to live as if it were true. The fact that most environmentalists quickly return to the comforts of capitalism after a brief fling with rural life or volunteer work among the poor does not deter them from continuing to endorse poverty and the simple life for others and proclaiming its joys. Their attempts to use others to prove their version of history can be as callous as experiments with human life.

Mark Plotkin, who is well known for his studies of herbal medicine, works for Conservation International in Latin America, where he has been urging natives to rely on traditional herbs. Many of these herbs turn out to have little effect on the diseases for which they are prescribed, however. Plotkin himself noted that several Indian tribes use a variety of antidotes for the paralyzing curare poison they put on arrowheads, but that when scientifically tested, these remedies do not work. Some traditional medicines even have effects as dangerous as the old European practices of applying leeches, bleeding, and purging. The result of Plotkin's medical approach will be history written in the bodies of dead brown men.

In Kenya and Tanzania, the great wildlife parks have taken 9,650 square miles of land from the Masai tribe. Among environmentalists, the Masai are considered "noble savages" and "natural ecologists," but they and their cattle are not allowed to be part of real wilderness. They must live apart from nature under what some writers have called "ecological apartheid." For many Masai today, their traditions are maintained mainly by money that "safari" tourists pay for photographs.

Francis Moore Lappe is best known for her *Diet for a Small Planet,* a counterculture cookbook, but she is also the environmental movement's guru on world nutrition and sustainable agriculture. According to Lappe and most environmentalists, the Green Revolution's new crops were created by technology, so they are "unnatural." Thus, they are not fit for the millions of people they have saved from famine. Lappe also criticizes the Green Revolution that has fed Asia for what it has not done; she writes, "It was a choice *not* to develop technology that was productive, labour-intensive, and independent of foreign input supply."[26] Lappe and others assume that labor-intensive agriculture is good for the world's poor. Anyone who has worked behind a horse-drawn plow, or turned the earth with a shovel or fork, or picked beans on a hot day would find this assumption not just naive, but cruel and arrogant. The vision of a world full of small farmers producing vegetables and grains with almost no chemicals for six billion people is quite pleasant unless you happen to be one of those who has to do the work and live off the proceeds. Environmentalists who readily spend other people's money just as readily condemn them to hard labor. They show little concern for the economic consequences of their programs on others.

> The developers promised some 1200 jobs. My job was to conduct public hearings, and people came from south Berkeley, which is the African American neighborhood, saying, "If we have this development, we'll have jobs." The environmentalists actually booed them; I felt like I was in the middle of Mississippi dealing with the Ku Klux Klan.[27]

Turning social and environmental decisions over to the environmental movement or the politicians in debt to the movement will have deadly results for both humanity and our environment. The environmental movement is right about its most general prescription—that for our own good, spiritually as well as materially, we must respect and cherish nature. Unfortunately, when the details emerge, Rousseau steps out from behind the curtain of history and tells us we must abandon civilization as we know it and assume the world view of primitive people, the "noble savages."

Rousseau's descendants, who propose unworldly solutions to worldly problems, now have more power than the descendants of Aristotle or Newton, who offer no world at all, only a way to see and think. Rousseau would have felt proud of his accomplishments listening to Vice President Gore describe the industrialized world as a dysfunctional family. He would certainly recognize his legacy in the words of Thomas Lovejoy, the science adviser to Interior Secretary Bruce Babbitt: "The planet is about to break out with fever. Indeed, it may already have. And we are the disease. We should be at war with ourselves and our lifestyles."[28] When a democracy is dominated by people who dislike its very origins, we have a crisis that deepens with each election.

CHAPTER 5

Searching for a New Sense of the Sacred

I never knew a people so eager to laugh, so dedicated to family. The only word that comes to mind is harmony.
—Kevin Costner as Lt. John Dunbar in *Dances with Wolves,*
on the Sioux

One day, a year before the breakup of the Soviet Union, I received a letter from a Russian friend. Its envelope bore one of those oversized and colorful stamps that the Communists printed in great variety to turn foreign mail into miniature propaganda posters. Instead of spaceships or declarations of friendship and peace, this stamp featured a mounted horseman rearing above the half-naked body of an Indian bleeding from a fatal chest wound. The image honored James Fenimore Cooper, the author of *The Last of the Mohicans.* When next in Russia, I discovered that nearly every Russian, from Moscow to Siberia, has read Cooper's novels.

Later, a young Slovak couple staying at my home in North Carolina asked if I had any books about the American Indian Vinetou. Most Eastern Europeans are exposed to the wisdom of Vinetou before they become teenagers. But Vinetou is not a product of the American West; rather, he was created by the German writer Karl May.

Such primitive heroes served two important functions under communism. For people living in a world of endless concrete apartment blocks, books about Indians, nature, and the New World provided escape. For Communist propagandists, the suffering of the wise and innocent Indian at the hands of greedy whites exposed the evil at the root of capitalism. According to Marx and Engels:

> The whole history of mankind (since the dissolution of primitive tribal society and the holding of land in common ownership) has been a history of class struggles, contests between exploiting and exploited, ruling and oppressed classes.[1]

Primitive people everywhere have become the saints of the environmental movement. In canonizing them, the movement has created an image that bears little resemblance to real primitive people. This reinvention of the primitive world is the heart of environmental theology and "science." Primitives have become the gold standard of the environmental movement. Against this standard it measures the values and achievements of our society. The historian Anna Bramwell has pointed out in her excellent history of twentieth-century environmentalism that the movement condemns industrial society for putting itself above, or "transcending," nature, but that like humans from time immemorial, environmentalists crave transcendence: "They seek the transcendental 'other' in primitive tribes."[2] Their "other" leads a life that somehow transcends without dominating. This blithe reinvention of other cultures says volumes about how far from the real world the movement has strayed and how willing it is to sacrifice other people's dignity for its cause.

Even the movement's objections to the words *primitive* and *Indian* reveal where its values lie. Environmentalists and politically correct liberals say that *primitive* implies ignorance, while in fact a Bushman or rain-forest hunter has a great store of knowledge, even if he cannot read, write, or drive a car. *Indian*, is a name imposed on people of the Americas by European explorers. All true. But I suspect these truths are not as important as trying to erase the achievements and advantages of Western culture.

I know the shortcomings of *primitive* and *Indian*, but the alternatives are awkward and imprecise. Therefore, I will use these imperfect expressions instead of the more imperfect "pre-European" or "pre-Columbian American" for *Indian*, and "primeval," "indigenous," "small-scale," or "land-based" for *primitive*. In English, *primitive* does carry a lot of baggage, and some of it is bigotry, but there is no better word with which to generalize, and listing all exceptions is unreadable. *Primitive* comes from *prime*, or first. I use it in that sense, and for its suggestion that the science and technology of such people are relatively simple.

INVENTING A GOOD EXAMPLE

If Karl May and James Fenimore Cooper reinvented Indians as mental and physical supermen, the environmental movement has ordained them as divine, or at least as the high priests of creation. In the opening scene of the 1992 film *The Last of the Mohicans,* Nathaniel, the white man brought up by Indians, shoots a magnificent elk in the primeval forest. Immediately, his Indian father, Chingachgook, kneels and says to the dead elk, "We're sorry to kill you, brother. We do honor to your courage, speed, and strength." Here in a nutshell is the supposed difference between the Indians who respect nature and those people whom Nathaniel later calls the "masses in Europe infected with the disease of greed." In reality, little documentation exists of the "thank-you dead animal ritual" as a widespread— or local—Indian practice before the environmental movement discovered it and modern Indians decided it was good public relations. The hunter's thanksgiving may have been natural to some primitive people who under hard circumstances were genuinely grateful, but does it put them any closer to nature than the American family that thanks God before each meal? One wonders how Plains Indians addressed individual animals when they ran hundreds of bison over a bluff in a single hunt or corralled a few dozen deer by burning the forest's underbrush.

Perhaps no story illustrates the distance between truth and fantasy better than that of Chief Seattle, leader of a small clan of Duwamish Indians on the shores of Washington's Puget Sound. In 1854, the sixty-eight-year-old chief, then frail and lame, negotiated a treaty with the territorial governor. At the time, Seattle made a speech. From the early 1970s these words have been quoted every Earth Day and they appear in Vice President Al Gore's book in 1992. Environmentalists have "quoted" the words of Seattle to shame Americans about what they have done to the continent. The vice president assures us that Seattle's vision has "survived numerous translations and retellings."

> How can you buy or sell the sky? The land? The idea is strange to us. . . . Every part of this earth is sacred to my people. Every shining pine needle, every sandy shore, every mist in the dark woods, every meadow, every humming insect. All are holy in the memory and experience of my people. . . . If we sell you our land, you must keep it apart and sacred, a place where man can go to taste the wind that is sweetened by the meadow flowers. Will you teach your children what we have taught our children? That the earth is our mother? . . . This we know: the earth does not belong to man, man belongs to the earth. All things are connected like the blood that unites us all.[3]

Unfortunately, Seattle's speech as so often quoted is phony, invented in 1971 by Professor Ted Perry for the ABC film *Home*. Vice President Gore, a former journalist who should know better and maybe did, is wrong. Chief Seattle was baptized a Roman Catholic and only freed his eight slaves after

Lincoln's Emancipation Proclamation. He made his reputation killing other Indians.

How the false speech has been used reveals how religious fervor corrodes important truths. To see the propaganda machine at work, let us compare a passage as rendered by Al Gore to how it exists in texts that he and his researchers should have, and may have, known. Gore quotes: "Every part of this earth is sacred to my people. Every shining pine needle, every sandy shore, every mist in the dark woods, every meadow, every humming insect. All are holy in the memory and experience of my people. . . ." (Gore ends with the ellipses, leaving off the reason for the ground being holy. He also does not give a source for his quotation.) The passage as quoted in two popular recent paperbacks and a museum publication reads like this: "Every part of this country is sacred to my people. Every hillside, every valley, every plain and grove has been hallowed by some fond memory or some sad experience of my tribe. Even the rocks which seem to lie dumb as they swelter in the sun . . . thrill with memories of past events connected with the fate of my people."[4]

Herein lies the critical difference. The latter passage, which is closest to what Seattle might have said, has no environmental message. The ground is sacred from a *human* point of view—because of human events and memories. The vice president, like most environmental writers, has turned Chief Seattle into a ventriloquist's puppet, a wooden Indian. Allowing history to be falsified, for whatever noble purpose, sets a bad precedent. It is a precedent that Germany, Italy, Russia, and China have lived to regret.

ON THE WARPATH AGAINST THE WEST

The stereotype of the noble savage turns primitive people into political pawns. It is a disturbing, even malicious, assault on both primitive culture and Western values, including those values that have allowed the environmental movement to be conceived, born, and raised. The basic message is that industrialized societies like ours only create problems. Speech after speech by government officials, environmentalists, and religious leaders cites the wonderful relationship that primitive people have with nature. By idealizing primitive people, environmentalists try to make other human beings *personae non gratae* in the natural world. Let us reconsider each of the assumptions that environmentalists make about primitive people.

PRIMITIVES IDENTIFY WITH PLANTS AND ANIMALS AS EQUALS

The preference of every species or group is its own survival. Human beings, primitive or industrialized, are no exception, although humans—and mainly those who are members of the highest, most industrialized civilizations—have become the sole example of creatures who also look out for the well-being of other creatures that have no necessary role in their survival. Yes, termites, farm fungi, and ants protect certain aphids, but the termites eat the mushrooms and the ants collect the sugars produced by the aphids. Humans tend roses. They care tenderly for dogs and cats, providing free health care and even burial benefits. They pass laws to save minnows and butterflies and to prevent other humans from acting cruelly to animals. Humans pay small fortunes to keep mountain landscapes pristine. Yet their survival does not depend on any of these activities.

Most primitive people not only put their species first, but unashamedly put their own nation or tribe first as well. They are brazenly ethnocentric and racist. Such prejudice should make Romantic anthropologists and liberals blush. But so often they admire in primitive people what they abhor in their neighbors.

The Lakota Sioux lawyer and writer Vine Deloria produces a long list of Indian tribal names and their meaning in his book *God Is Red*. Almost all are ethnocentric. A few examples:

Ainanbai (Maine's Abnaki)	men or people
Lenni Lenape (New Jersey's Delawares)	true men
taneks aya (Mississippi's Biloxi)	first people
ani yn wiya (Georgia's Cherokee)	real people
Totcangara (Wisconsin's Winnebagos)	people of the real speech
Kiowa (Oklahoma)	principal people
Hum-a-luh (Skagit of Washington)	the people[5]

Can we conclude that for the Indian, all animals were equal with the exception of other human animals? The Orwellian nature of remanufactured Indian versions of equality comes across loud and clear in *Dances with Wolves*. When filming, the director, Kevin Costner, consulted with the Sioux about how to portray the different Indian tribes. As a result, the Sioux's historical enemies, the Pawnee, are presented as irredeemably cruel, bloodthirsty, their Mohawk haircuts as symbolic as the Western villain's black hat.

If living in simple family groups close to unspoiled nature is supposed to produce kindness toward other creatures, most primitive people have not gotten the message. Allen Johnson, an anthropologist at the University of California, lived with the Machiguenga in Peru's Amazon basin. He observes:

> The Machiguenga view of nature is self-centered and ambivalent. When nature is bountiful, it is loved; when it is unpleasant or dangerous, it is despised. What nature offers is taken in the most convenient manner; what nature threatens is avoided.[6]

The Machiguenga are usually kind to each other and to household pets, but Johnson saw men, women, and children torture captured animals for the amusement of onlookers, "with both the torturer and the group laughing riotously at the pitiful cries of the victim."[7] A children's sport among the Yanomamo of Venezuela and Brazil is tying a lizard to a string anchored to a stick. They then happily chase it around and around, shooting arrows into it. The same behavior in Europe or America can bring a stiff fine or a jail sentence.

If kindness and respect for animals are a sign of being close to nature, it is far from clear whether primitive people or people who drive cars and watch television feel closer. I suspect that environmentalists have popularized the story that Indians thanked just-killed game in order to convince people that animals accept being killed by primitives but not by moderns. In the environmentalist guru Gary Snyder's opinion:

> Other beings do not mind being killed and eaten as food, but they expect us to say please, and thank you, and they hate to see themselves wasted. . . . In their practice of killing and eating with gentleness and thanks, the primary peoples are our teachers.[8]

I wonder if Gary Snyder would mind being killed and eaten by a tiger if it would say please and thank you.

PRIMITIVES HOLD LAND TO BE SACRED

The idea of the sacred landscape may be yet another European-American invention. The distinction between nature for Westerners and for primitives is supposed to be the presence of the divine. Because the divine presence is a good defense against development, it is frequently invoked in modern-day legislative hearings and fundraising campaigns for environmental groups. Indians, ancient and modern, never had so many holy places until the Americans decided that Indians would become the symbol of pristine nature. Chief Seattle's falsified words are a good example. Holy

landscapes were put into his speech by a white professor and canonized as
scripture by environmentalists. But those places were holy to Chief Seattle
only in the manner of battlefields such as Bull Run, Verdun, or Porkchop
Hill, or historical sites such as Plymouth Rock or Williamsburg.

If it is not history or battles that hallow nature, sacredness for primitive
people may find its source in myths about gods and spirits, metaphysical
beings that think and behave suspiciously like human beings. Some primitive gods and spirits are kind and merciful. Most are not. Often they
behave like the worst humans. They are fickle, deceptive, vicious, and dictatorial.

The Yanomamo mentioned earlier believe that the world is like a four-
layer cake. The top layer is generally empty. The second layer, just above
the blue dome of the sky, is a mirror image of the earth, with people there
doing everything the Yanomamo do below. Our world is a piece of this
upper world that broke off and fell. Below us is a barren world with one village of spirit people. They have no jungle and no game, so they send their
spirits up to earth to capture and eat the souls of children.

If behavior is any clue to religious belief, primitive people consider
most of nature to be no more sacred than do Europeans, Americans, or
Japanese. The answer to saving the environments in which primitive people still live is not in hoping that they will continue to be themselves, but in
hoping that if they really love nature, they will recognize that its salvation
lies not with the gods but with scientific management.

Several Pacific Northwest Indian tribes understand this. In the magnificent evergreen rain forests, the Quinault Nation has set up a modern
forestry program. On Wyoming's Wind River Reservation, the Shoshone
and Arapaho have set up their own Environmental Quality Commission.
The policies made by the commission are enforced by eight armed wildlife
wardens on horseback. In place of the traditional free hunting that devastated wildlife populations, they are enforcing hunting seasons. And to give
wildlife a boost, they have implemented an active environmental restoration program.

PRIMITIVES DO NOT BELIEVE IN LAND OWNERSHIP

For environmentalists who believe that the greed of industrial countries
will destroy the world, the main target must be private property and the
capitalist system it supports. However, industrialized socialism and communism have made bad names for themselves, so that leaves only primitive
communalism as an alternative. Since the late 1950s, scholars caught in the

passion for rethinking society have proposed that primitive people lived in a primitive communism, sharing everything important. Natural resources and land especially were supposedly held in common by the tribe, partly because the resources were sacred, partly because no one had been corrupted by greed.

Environmentalists think that the absence of deeds and registered titles proves that private property did not exist. Indians, of course, had no archives or records except art and oral history. However, even in these there is evidence of material possessiveness and property claims. On the East Coast, tribes not much larger than extended families owned shellfish beds and fishing areas, not to mention garden plots and hunting grounds.

The myth of primitive communalism has even shaped U.S. foreign policy. In Kazakhstan, background material given to U.S. advisers on privatization often contains articles and statements warning that privatizing land is not feasible because of the Kazakh nomad traditions that have created a strong suspicion of anyone trying to own land. The fact is that for a thousand years Kazakh families had very strong land ownership rights and fought fiercely to protect them. Colgate University political scientist Martha Brill Olcott points out in her exhaustive history, *The Kazakhs*, that bias against property was a Communist creation. "The development strategy chosen—that is, collectivization and nationalization of almost all property in the rural sector—was especially devastating to the Kazakhs; traditional Kazakh culture defined a man through the animals he owned, making private ownership of livestock almost the definition of what it was to be a Kazakh."[9]

As individuals, families, and groups, primitive people staked out ownership claims, and they waged endless warfare to expand old territories or acquire new ones.

PRIMITIVES DO NOT TREAT NATURAL RESOURCES AS COMMODITIES

Beyond their simple needs for food, clothing, and shelter, we are supposed to believe that primitive people did not care much for luxury or material consumption. When they did kill birds or cut trees to enlarge their houses, embellish their rituals, or improve their transportation, we are to assume that they did so with great respect for nature. The accumulation of wealth was not a high priority.

The first test of anyone's values is to offer them choices. What do primitive people do when offered guns, metal cookware, central heating and air conditioning, movies, cameras, alcohol, cars, laboratories, cowboy boots,

Gore-Tex parkas, snowmobiles, and electric guitars? No matter what the color of the writer, history records only a few instances when primitive people did not choose and try to maintain a consumer lifestyle:

> They seem to appreciate instantly the efficacy of a steel machete, ax, or cooking pot. It is love at first sight, and the desire to possess such objects is absolute. There are accounts of Indian groups or individuals who have turned their backs on manufactured trade goods, but such people are the exception.[10]

The facts say that primitive people used natural resources as selfishly as any modern culture. They simply lacked the tools and incentives to do as much damage. Nothing in their religion or science stopped them. In fact, with a little encouragement, they destroyed natural resources with greater abandon and ignorance than more scientifically advanced cultures.

When the first Asians streamed across the land bridge now under the Bering Sea, North America's forests, plains, and deserts still had large populations of mastodons, camels, giant beavers, wild horses, and three-ton ground sloths. Like all hunting cultures, the new Americans hunted out one place and moved on. Experts estimate that a small band of hunters could exterminate all the large animals near one settlement in as little as ten years. The immigrants probably multiplied rapidly in the rich environment. America seemed as endless to them as to eighteenth- and nineteenth-century lumbermen.

Far from being natural ecologists, primitive people seem to have had little understanding of how many animals it takes to maintain a viable population. In large areas such as North and South America, Australia, or Siberia, primitive people could survive their ignorance by moving about. Sometimes the land recovered after they moved on. By the time Columbus arrived, however, Indians were knocking elbows with one another. Depending on circumstance and custom, populations were kept in check by warfare, a life span limited by lack of medicine, selective human sacrifice, and killing unwanted babies.

When Europeans arrived on the eastern seaboard, they were surprised by the relative scarcity of deer and other game, but quite pleased with the great trees and open woodlands. The woodlands were often free of scrub because the Indians had burned the undergrowth for better hunting. Obviously, they cared little for the "uneconomic" species sustained by the forest understory. In the end, the practice was self-defeating because deer and many small animals depend on brush for food and shelter. Biologists generally agree that in many areas today, despite rifle hunting, deer are more plentiful than they were when Columbus arrived.

Most "natural" native cultures lived in high-risk, high-misery situations. Their populations expanded until some limit of the environment stopped them. Then they went to war for more food and production materials (such as wood, clay, and stone). If they had no competition, they devastated their environment without regard for the future.

The customs that led primitive people to exploit nature were often as egocentric and frivolous as our own. Some of Hawaii's extinct birds ended up in beautiful feather capes. The Pilaga' of Argentina's northern Formosa Province include in their elaborate bride-seduction rituals small packages "containing the stems of wild plants. . . . The bundles also contain parts of song birds—feathers, eyes, hearts—or other animals reputed to sing beautifully—some kinds of frogs and crickets."[11]

Most humans are obsessed with their material needs and their image as expressed by possessions—whether a feather cape or a Mercedes-Benz, a jar of caviar or a rack of bison meat, a closet full of shoes or a tundra full of reindeer.

Primitives Are True Ecologists

Primitive people knew little about extinction, and when they had the will and the way, they erased many species forever. A changing climate may have challenged the mammoths, mastodons, giant sloths, North American horses, and other "megafauna" that flourished at the end of the last ice age, but almost certainly primitive people delivered the final blow. Archeologists working on Cyprus recently uncovered yet more proof that primitive values or knowledge did not blink at extinguishing rare species: in a cave shelter with thousands of stone tools, they found a huge deposit of some 250,000 bones. Most of them were from the extinct pygmy hippopotamus.[12]

One of the great myths about the primitives is that they were ecological farmers in tune with the forces of the universe. Primitive people invented agriculture and animal husbandry. Agriculture is the single most destructive force in the natural world, far more destructive than shopping malls, superhighways, hydroelectric projects, or nuclear power plants. The purpose of almost all agriculture, primitive or modern, is to define a piece of land, eliminate all natural plants and animals, and grow only one crop, whose seeds, leaves, fruits, or stalks are consumed almost solely by the human species.

In some cases, irrational fear of sacred nature led to bloodletting among humans. Mayans sacrificed human beings to keep the power of the universe fertile: "As maize cannot seed itself without the active intervention of human beings, so the cosmos required sacrificial blood to maintain life," that is, the cosmic order depended on human action.[13] Among the other practices that primitive people have used to "preserve" the ecological balance are infanticide, slavery, and the execution of unwanted citizens, especially women.

Such practices were part of a cultural ecology that became institutionalized and legalized. To sanctify primitive people on the basis of their virtues alone is like beatifying Mussolini because he made Italian trains run on time or nominating Hitler for a Nobel Peace Prize because he signed a nonaggression pact with Stalin.

All human groups tend to know what is necessary for survival or they wouldn't be here long. And they tend to know those things that interest and entertain them. That is true for our own knowledge of nature. Surfers know waves, photographers know light, plumbers know water physics, gardeners know flowers, and so on.

In some cases, the gods of primitive people actually separate them from nature. The Indians living near Mount St. Helens (the Salish to the north and west, Cowlitz to the southwest, and Klickitats to the southeast) believed that in it dwelled Loo-wit (Keeper of the Fire), Lawelatla (One from Whom Smoke Comes), or Tah-one-clah (Fire Mountain). John Staps, a Klickitat who helped a party of whites climb to the summit in 1860, was later ashamed of his involvement. He said that Indians did not venture onto the mountain except to prove their bravery by walking in the high meadows in the presence of the spirit that ruled the mountain. Young Klickitats who did this were admired by the elders and received into the secret councils as a brave, no longer a *tenas man* (adolescent).

Fear of Mount St. Helens may have protected the Indians from its occasional explosions or gas clouds, and it may have left the slopes as a kind of game preserve to replenish herds wiped out by Indian hunters. In this respect, the superstition is not unlike the broad-sweeping prohibitions that environmentalists want in order to control the dangers of chemicals.

At other times, such beliefs are counterproductive and even dangerous. The *Amicus Journal,* published by the Natural Resources Defense Council, describes the Hopi Indians' dilemma about natural resource development and modernizing their reservation. In a sympathetic article, the Hopis tell the writer that if the place where they find snakes for ceremonies is disturbed by a road, "all of the Hopi Way will be dangerously out of balance." Another Hopi leader says that because of the power lines that have crossed the reservation, they "don't get rain anymore. The whole earth is out of balance."[14]

In the world of Peru's Yuqu, spirits threatened from every tree and animal in the rain forest. If someone in their upper ranks died, he had to be accompanied by the spirit of another human being, so the Yuqus killed one of their slaves—another Yuqu with the misfortune to be born into the bottom of society. The Yuqus also cut themselves off from their own history by a taboo against ever speaking the name of a dead person, lest that person's spirit rise and attack.

People who accept the myth of primitive harmony with nature also argue that primitive people make kind and efficient use of land. In the struggle over Indian land claims in Brazil, some interesting numbers have emerged. The Kayapo of northern Brazil were aided in their quest for the 19,066-square-mile Menkragnoti Reserve by the rock star Sting, who raised $2 million for their cause. Perhaps the Kayapo are entitled to this private property, whether they need it or not; lots of people own idle land, and a few fortunate wealthy people own forests and plantations of more than one thousand acres. However, very few people anywhere in the world own more than a single square mile. Yet the Kayapo have claimed 9.53 square miles for every man, woman, and child. If that is what the Kayapo need to survive, they are hardly very efficient at using the supposedly rich lands of the Amazon basin.

The monopoly of large tracts of land is an evil that environmentalists generally charge against white capitalists; why have environmentalists never asked how the Kayapo and other "native" people acquired their territory? Around the world, the acquisition of uninhabited lands generally ended some 5,000 to 10,000 years ago, as the great post–Ice Age migrations reached every mountain, valley, island, and plain of the world that was not permanently covered in ice or snow. Oral and archeological history suggest that most original claimants were violently displaced over time. If we were to honor only those land titles that were claimed peacefully, the Iroquois would lose their New York lands and the Sioux, their lands in the Black Hills and plains. The Kayapo would lose their vast holdings in Brazil. Perhaps only the Australian aborigines could claim that their ancestors had displaced no one else.

Despite the great injustices and swindles that conquerors have inflicted on native people, the natives have generally fared better than the people they themselves conquered. Many Indians have come to enjoy a land tenure more secure than any their ancestors dreamed of. Some examples:

On the Venezuelan border, Brazilian authorities finally decided in 1990 to help the Yanomamo protect their property rights by removing some 45,000 trespassing gold miners. The Yanomamo were left with some 3.9 square miles (2,496 acres) each, land that includes rich gold deposits.

In the 1980s, some 39,000 nonwhites living in Canada's vast northern lands were given title to 260,000 square miles, about 6.6 square miles (4,224 acres) for each person. The agreement promised them a strong voice in development proposals for another 1.1 million square miles, about 30 percent of Canada.

In 1988, 6,500 Yukon Indians received 16,000 square miles, about 2.5 square miles (1,580 acres) each, plus $200 million ($12,121 for each person).

Reservations and private Indian ownership in South Dakota claim 87 acres for each adult and child.

In Wyoming, each Shoshone and Arapaho living on or near a reservation has some 382 acres.

In New Mexico, Navajo, Apache, and Zuni living on and off reservations hold title to an average of 62 acres each.

Sometimes the Western land-use practices that environmentalists condemn become models of ecological sensitivity when practiced by people in loin cloths, *huipiles*, or nomadic robes. The Rutgers University philosophy professor David Ehrenfeld praises ancient Middle Eastern farmers who destroyed hillside ecology to grow valley crops:

Some of the finest desert agriculture of all times was practiced by the Nabateans, who lived in Israel's Negev Desert 2,000 years ago. Growing luxuriant crops in places that receive three to four inches of rain in a good year, the Nabateans worked miracles. They placed their farms in the valleys of the hilly desert, lining the hills with runoff channels to convey every possible drop of rain to the farms below. The runoff channels terminated in a complex water distribution network of irrigation canals and weirs.[15]

Ehrenfeld goes on to describe how a professor of botany at Hebrew University discovered the reason for thousands of piles of stones in the region: "[He] found that raking the stones out of the desert soil greatly increased the flow of rainwater to the farms below."[16]

If the Nabateans had been modern farmers, Ehrenfeld might be aghast at how they destroyed the natural ecosystem's normal water distribution to hillsides and the water flow in the valleys, but instead he calls this "the ultimate demonstration of the Nabateans' genius." He concludes with a statement whose import he cannot possibly understand: "Like most other places on this living planet, the desert is what one makes of it."[17] For environmentalists, primitive people are smart and sensitive when they make something of the environment, but not us. No matter what they do, primitive people are true ecologists:

> Despite evidence to the contrary, indigenous people continue to be credited with a natural respect for ecology and a commitment to sustainable methods of resource use under all circumstances. Some Indian groups, reading of the qualities attributed to them by Europeans, have begun to give themselves the same credit. . . . The assertion that as Indians these people will be ecologically noble stewards, though unproven, is a trump card in the current world of conservation sensitivities.[18]

THE SAVAGE NOBLE

Despite a lot of conjecture, historians and archeologists have provided a usable picture of pre-European cultures on all continents, and for the twentieth century, when documentation has been more methodical and detailed, we have good descriptions of contemporary primitive people. Past and present, they bear little resemblance to the ecological savants of environmental literature.

The environmentalists' creation is not a human being, but a symbol, and it is a symbol that distills a strong prejudice against Western values. Just as Christians have an antichrist, the noble savage is an antiscientist. Just as the antichrist often appears in a form that mimics Christ, the antiscientist often appears to mimic science. We are told that primitive people's knowledge of medicine, agriculture, and animal behavior far exceeds ours, and that they possess powers over nature unexplained by science.

The ideal of the noble savage has legitimized its modern counterpart—the savage noble. The obvious example is the rock musician—wild looking, passionate, and above all, ignorant of what we call science and civilization. Motley Crüe, Nirvana, Sinéad O'Connor, Madonna, and scores of other teenage idols are innocent savage nobles, the adolescent's mercenary soldiers striking back at a cruel, cold society that their fans believe is bent on killing all that is good and natural in human nature.

The environmental movement does not sing the lyrics of rock and roll, but it has absorbed much of the distrust of science and has idealized primitive innocence, which is too often a kind phrase for ignorance. Of course, it is impolite and politically incorrect to call any ethnic or social group ignorant, unless the group is within your own culture. Fundamentalists of the rain forest, desert, or tundra are never ignorant, but fundamentalists of the Moral Majority or Petaluma are. To consult the moon and the stars about the future from the remote Himalayas or a Mayan pyramid is mystical communion, but when Nancy Reagan did it from the White House, it was stupidity. This is a double standard whose rules damn civilized people no matter what they believe, unless they strip to loin cloths and sandals, or at least shampoo with jojoba, eat Rain Forest Crunch, and use recycled toilet paper. Or perhaps unless they join an environmental group.

Among environmentalists, a fleeting and distant knowledge of primitive people allows them to project onto those people, living or dead, the ideal of generous communal human beings living in, and loving, their friendly deserts and rain forests. Few environmentalists live in the real world of primitive people. It's a nice place to visit, but who would want to live there? Almost no one who spends months or years with primitive people idealizes them. The very few hard-core primitivists in the environmental movement seem to sustain themselves in a bubble of passionate ideology—my people, right or wrong! Or they are so absorbed in their particular field—finding medicinal plants or spiritual revelation—that they are shielded from reality by their own self-centeredness.

Primitive people lived, as a few still do, in a frightening world of unpredictable spirits and prohibitions as broad as their ignorance of nature's full complexity. The fear of nature's gods and the development of broad superstitions may have been clumsy but necessary adaptations to the environment and protections for it, but what produces security under stable conditions becomes handcuffs and leg irons in times of change and crisis. Changes in climate, new diseases, swings in animal populations, and the sudden appearance of other human cultures devastated primitive societies; while they may have lived "happily," they also lived short and unpredictable lives.

CHAPTER 6

Winning the Public Away from Science

Most people have eyes and can, therefore, see. However, few people have the ability to reason. Therefore, appearances are everything.

—Niccolò Machiavelli, *The Prince*

Popular horse-racing tabloids are much more scientific than the environmental press. People who write the racing sheets know the uncertainty and risk in their predictions. News on the environment features experts who claim to be almost entirely certain about their predictions. Claiming to be certain, especially about a future disaster, can sell books and raise funds. Even when the writers are not sure of the facts, they will place a heavy bet, usually with someone else's money. In the case of developing countries, they will even wager someone else's comfort or life span. So do politicians + the press

The purpose of environmental journalism is not to convey the truth about nature or the impact of civilization. Rather, its goal is to sell the myths of the movement. The environmental movement appeals for converts in much the same way that Pascal argued for people to believe in God. Although the seventeenth-century French philosopher doubted the existence of God, he proposed that a prudent person should bet on the Almighty. After all, he argued, if there is a God, the believer has everything to gain and the nonbeliever, everything to lose. But if God does not exist, the believer and nonbeliever go to the same fate.

The environmental movement's grip on the media relies on information that trickles down from a point where great generalities are manufactured. At the grass-roots level, anecdotal information provides the "proof." Global warming is a fine example of the environmental trickle-down theory. In the 1980s, when computer models predicted that as carbon dioxide levels rise, the atmosphere will warm, anecdotes poured in about local heat waves and droughts. When the summers of 1987 and 1988 brought unusually high heat, Al Gore, then a senator, joined environmental groups everywhere in suggesting that this could be global warming in action. However, not a single credible scientist believed that the North American summers proved anything—except that if you don't like the weather, wait a few years.

In 1984, environmentalists seized a huge media opportunity when Lester Brown's Worldwatch Institute published its first annual state of the world report. Nobody else was publishing such material. Brown began the first volume with a flood of highly selective tables and charts and highly biased analysis. In a vacuum of other readable reports, *State of the World* became a popular reference book for journalists, especially those sympathetic to the environmental movement. Each year's report often sells more than 100,000 copies (more than many best-sellers).

In 1986, *Time* magazine wrote that "Brown's presumptuous 263 page volume may be studied more intently by more people in more countries than Reagan's [state of the union] address."[1] The MacArthur Foundation, always ready to finance apostles of the apocalypse, awarded Brown a $250,000 "genius grant." The Worldwatch Institute's budget is well over $1 million a year, almost entirely slated for generating information for the environmental movement. But to read the annual reports three to five years after they appear, the figures and predictions become useless, or are exposed by later events as simply fraudulent. For example, Brown's 1989 report touched off a huge campaign to stop logging in the Amazon by citing fuzzy satellite data combined with guesses about the number of fires burning. He warned readers that "satellite data show that 8 million hectares [20 million acres] were cleared in 1987 alone, 5 to 6 million more than were thought to have been cleared annually in the early eighties."[2] Brown never revealed that his figures were based on a Brazilian study that has been refuted by many competent scientists. By 1990, improved interpretations would slash Brown's estimate by an area the size of Zimbabwe. In 1993, better data led scientists to estimate that the rate of annual clearing was actually 2.1 million hectares between 1978 and 1989, falling to 1.4 million hectares between 1989 and 1990.[3] In his 1989 report, Brown had claimed almost 400 percent more deforestation than actually existed. Despite his reliance on questionable data, the popular press continues to give Brown's predictions and annual reports reverent coverage.

The environmental movement has been a blessing to the media. Scientists have traditionally remained too cautious and modest to provide the excitement the media needs to get readers, listeners, and viewers. Environmentalists, using the language of science, seem to provide the easy answers and the drama. Einstein once compared the pursuit of atomic physics to following the tracks of a great beast across the universe. Scientists are still following such tracks, but with less hope of seeing the beast. They have realized that nature and the universe are far too complex to be fully understood by the human mind. From a religious point of view, this may be the same as admitting we will never see the face of God. Through the environmental movement, however, the popular press, radio, and television are always seeing the face of God. For them, scientific uncertainty seldom exists, or if it does, it is kept in the back closet like a crazy relative.

Environmental groups will sacrifice even their own goals to remain politically correct. One of the most difficult stories for the environmental movement is the fate of the elephant in Africa. The worldwide ban on elephant ivory is supposed to save the elephant. Ironically, however, those countries that enforce the ban continue to lose elephants to poachers, while in Zimbabwe and Botswana, where ivory harvesting is legal, elephant herds have been increasing. This is because local people want to protect a valuable source of income, an entirely rational response. However, fundraisers for World Wildlife Fund International (WWF) convinced the fund's experts not to use this information, since maintaining the ban is better for fundraising. England's Prince Philip, WWF's president, wrote in a confidential memo:

> The awkward problem for WWF is the massive popular support for a total trade ban regardless of its chances of success or its consequences. The ban is a typical "knee-jerk" reaction by the "greens" and it is not going to be easy to get rational measures adopted for the control of poaching and for limiting demand to sustainable supply.[4]

At the next higher level of environmental hysteria are public radio and television. Financed primarily by large foundations, environmental coverage is often quite different from higher-quality scientific programs. The prestigious show "Frontline," for instance, launched a frontal attack on agricultural chemicals to celebrate the thirtieth anniversary of Rachel Carson's *The Silent Spring*. To demonstrate the dangers of farm chemicals, the earnest Bill Moyers cited the "McFarland cancer cluster" in California. In fact, no epidemiologists have been able to link those cancers to farm chemicals. Every farmer knows the dangers of heavy doses of pesticides, and farmers and farm children generally enjoy lower cancer rates than the rest of the population, a fact that "Frontline" could not or would not use. If small doses of agricultural pesticides do cause health problems, the problems should appear in farm families first.

Few news or documentary pieces ever report important advances in technological management of environmental quality. For example, in the same period that the Carson anniversary show pretended to be competent science reporting, "Frontline" turned down several proposals by accomplished producers for a show on new methods for assessing the risks from toxic chemicals.

When scientific uncertainty cannot be avoided, everybody finds a way to take advantage of it. For many political conservatives, it becomes the reason to do nothing until science fills in all but the tiniest gaps. Environmentalists, however, will fill in large gaps with conjecture and often with outright fiction. If in the contest of evidence they find themselves surrounded and in short supply of facts, they will dig in waiting for new supplies. Air pollution has been an excellent example. No one likes dirty air except perhaps cigarette smokers. But getting people to look beyond the costs and start a quick clean-up requires certainty of imminent disaster.

In the 1970s, people such as the astronomer Carl Sagan and the climatologist Stephen Schneider warned that atmospheric pollution would cause global cooling. In an essay entitled "Eco-Catastrophe!" in Friends of the Earth's 1970 *Environmental Handbook,* Paul Ehrlich looked into the future and imagined a 1977 issue of *Science* magazine announcing "that incident solar radiation had been so reduced by worldwide air pollution that serious effects on the world's vegetation could be expected."[5] To reach this wrong conclusion, the doomsayers filled in huge gaps in climate research with pure guesswork.

The evidence in 1993 for global warming may have been more sophisticated than the 1970 evidence for global cooling, but it was hardly more decisive. Global warming exists only in theory and in the realm of quite limited computer models. No competent scientist claims to have measured on a thermometer, or any other meter, an actual increase in average global temperature. The best data we have on the overall global temperature come from NASA's satellites. An analysis provided by NASA and the Earth System Science Lab at the University of Alabama demonstrates that from 1979 to 1993 there was no significant trend toward either cooler or warmer temperatures. From 1988 to 1991, when Al Gore was writing his book, global temperatures were warmer than average. From 1984 to 1987, and in 1992 and 1994, they were cooler than average. Nevertheless, environmental groups and politicians have filled in the uncertainty gap with information that wins financial support and votes. Perhaps by the time these words are published, the gap will have narrowed, and Ehrlich will finally have bet on a winning prediction. But for now, his bet is on winning rhetoric.

In the meantime, the "certainty" of global warming provides the focal point for Bill McKibben's *The End of Nature* and Al Gore's *Earth in the Balance*. Even after Gore's election he continued to exploit false evidence for global warming. In 1993, he asked an environmental conference in Kentucky if it were just a coincidence that carbon dioxide has gone up along with temperatures in the last fifteen years.

Dominating media attention is more important for the environmental movement than getting the story straight. The Conservation Fund's 1989 survey of environmental organizations in forty-nine states makes the point. The fund asked each organization for its most important strategies. In their responses, nothing approached the importance of using the media to change public opinion. Seventy-two percent of the environmental leaders rated the media strategy as "very important" or "most important." Monitoring government agencies came in a distant second, with 51 percent rating it "very high." When the fund asked about the highest priority within the organization, only 3 percent said "hiring professional researchers." So where do they get their most valuable information? Among the possible sources proposed as extremely useful by the fund, more than half of the environmental leaders chose talking to other environmental leaders. Only 18 percent listed "discussions with leading thinkers in conservation history," and 24 percent chose "a fellowship in natural resource management." When evaluating newly hired staff, 57 percent of the environmental leaders surveyed rated their employees as excellent in interpersonal communication, but when it came to scientific knowledge, only 22 percent gave their own staff members an excellent rating. Fewer than half of the environmental leaders had a bachelor's degree in one of the sciences.[6]

The Sierra Club's 1970 *Ecotactics* laid down the guidelines for the movement's domination of the media. It was called "advocacy journalism" and its intent was to "destroy the dangerous anonymity afforded by objectivity or 'balance.'"[7] The chapter "Getting into Print" ends by noting with approval that "at the meeting of the U.S. Student Press Association in Boulder, just before Woodstock, student editors began realizing the necessity of politicizing their papers."[8]

Advocacy journalism on environmental issues has since spread to most major television and print outlets. When the Smithsonian Institution sponsored a conference for media people in 1989—called "Global Environment: Are We Overreacting?"—the answer was built into the program. Panels were loaded with speakers like Ehrlich and Brown. There was no dissent.

At the conference, the *Wall Street Journal* editorial writer David Brooks reported that Charles Alexander of *Time* said, "As the science editor of *Time* I would freely admit that on this issue we have crossed the boundary from news reporting to advocacy." NBC's Andrea Mitchell agreed: "Clearly the networks have made that decision now, where you'd have to call it advocacy." The *Washington Post* executive editor Ben Bradlee, noting that C-Span was taping the conference, warned about the dangers of being so public about the media abandoning objective reporting. He did not object to the practice, but he was concerned about how it results in "a whole kooky constituency to respond to, which you can waste a lot of time on."[9]

Environmental journalism runs the gamut from sleazy to brilliant, but the advocacy journalism promoted in the 1970s has taken over environmental magazines and newsletters. It has become the main source of information for school books and curricula, college courses on environmental issues, and citizen action groups.

Some writers and professors have legitimized advocacy journalism by claiming that objectivity does not exist. Perfection in any discipline, as Plato observed over 2,000 years ago, does not exist in the human world, only shadows of perfection. Yet the impossibility of purely objective reporting is a miserable excuse for not trying. Besides, if writers and environmentalists insist on undermining science by mocking *its* objectivity, they could destroy the only means we have for understanding what we want to save and whether our efforts are succeeding:

> Those who manipulate science as a tool of persuasion do not respect the sanctity of science. Nor do they respect the sanctity of an individual's right to self-determination. Their belief is that when given the truth people do not have the ability to make the "correct" decisions, so they must be lied to instead. Only the environmentalist manipulators of science have the ability to come to "correct" decisions.[10]

Well-reasoned and fairly argued writing exists, but it is generally ignored or picked over for partisan goodies as if it were nothing more than flea market trash and treasure.

The controversy over dioxin is a particularly good example of how advocates get locked into one point of view and then refuse to adapt to contradictory evidence. Dioxin, a by-product of pesticides, herbicides, and other industrial products, first became famous when the U.S. military sprayed massive amounts of Agent Orange on the forests of Vietnam to expose guerrillas in hiding. Many veterans blamed later health problems on the dioxin in Agent Orange. This became proof for the environmentalists' bias against synthetic chemicals, and they rushed to celebrate.

However, epidemiological studies turned up no direct link, or even a pattern of illnesses, that could be caused by dioxin. Studies of people exposed to dioxin after a chemical plant explosion in Seveso, Italy, in 1976 also showed no pattern, except pregnant women deliberately aborting their babies in fear of predicted birth defects. The lack of evidence did not deter environmental journalists at the time. Luckily for them, their fabrications were recently given substance by a newly emerging number of soft-tissue cancers in the exposed populations.

Despite such chance rescues, endless hollow claims of disaster in the 1970s and 1980s have invited a conservative backlash. Environmentalists have been wrong so many times that conservatives have stopped listening. They glibly dismissed the discoveries in 1992 that dioxin-like chemicals bind to important sites within animal cells (including human cells). There the chemicals exert their influence by slightly depressing the immune system and changing the growth pattern of embryonic cells and sexual development. Many conservatives assume that toxic chemicals are just a liberal bogeyman. The conservative mistake has been to discount any prediction of environmental disaster solely because an environmentalist makes it.

Edward Krug, who calls himself a recovering environmentalist, started his career as an acid rain expert and participated in the National Acid Precipitation Assessment. In Connecticut and at the national level, he discovered "how environmental goals and standards are currently established." He found that claims that 44 to 100 percent of Connecticut's lakes would die from acid rain were based on figures that were "entirely fabricated."[11]

Krug discovered similar baseless figures from more respected sources. In 1980, EPA claimed that "the average Northeastern lake had been acidified 100-fold over the past 40 years as a result of acid rain."[12] The next year, the National Academy of Sciences said that by 1990 acidification would increase by the same amount as during the past forty years. Krug says he was happy to find that the bad news was false news. All of these claims were scientifically unsubstantiated. Nevertheless, Krug watched as the media across the country turned acid rain into the cause of a new "silent spring." Before he announced his good news to environmental leaders, he thought that "everyone would breathe a sigh of relief in knowing that thousands of lakes and millions of acres of forest were not being sterilized by acid rain. To my shock, instead of thanking me, environmentalists fell on me like a ton of bricks." Krug concluded that "environmentalism is not a science but a socioeconomic revolution. Facts are ignored and other 'facts' are manufactured to support the deception of self and others."[13]

The simplest form of advocacy journalism pretends there is only one side to an issue. A few years ago, *Audubon* ran a series of articles on hunting by a hunter-conservationist. The outcry from readers put an end to using the magazine for serious exploration of other points of view on hot topics.

Even when environmentalists try to foster open discussion, they often get cold feet. In 1991, the Sierra Club invited the toxicologist Bruce Ames to write a piece for the centennial issue of *Sierra*. The editor-in-chief Jonathan King wrote:

> We plan to present the thoughts of a wide range of individuals whose knowledge, experience, and creativity qualify them to answer the following question: Given that our planet faces a future that appears bleak if not terrifying, what steps would you take to ensure our survival for the next 100 years? . . . please be as specific as possible.[14]

Ames is famous and respected for creating the standard tests used to determine the cancer-causing powers of many chemicals. He submitted an article to *Sierra* about how fact should be separated from fantasy in managing environmental risks.[15] Among other points, Ames noted that "life expectancy continues to increase in industrialized countries" and the age-adjusted rates for all cancers excluding lung cancer have remained steady or decreased since 1950. He also noted that 99.9 percent of all the chemicals that humans ingest are manufactured by nature, including some that act like dioxin. This figure includes some five thousand to ten thousand natural pesticides and their breakdown products. For example, Ames said, potatoes "naturally contain a fat soluble neurotoxin detectable in the blood stream of all potato eaters." At high levels, this compound causes birth defects in rodents. From a toxicologist's point of view, the choice is not between nature and industry. It is between what is more or less beneficial: if we want an adequate supply of affordable food such as fruits that help prevent cancer, we also have to ingest chemicals. Ames's preferred policy would be to assess the real risks of using synthetic chemicals and weigh them against the benefits. King rejected the article, citing stylistic reasons: "We found the focus on pollution and pesticides far narrower than the sweeping prescriptions other contributors were making."[16] Apparently Ames was not terrified enough about the earth's future, or the Sierra Club was too terrified by his science.

BAD JOURNALISM COSTS MONEY, MAYBE LIVES

Demos of megacorporation marketeering practices only 1 side of story told, as impending doom.

Magazines published by the Sierra Club, Wildlife Federation, Natural Resources Defense Council, Wilderness Society, and Audubon Society reach more than seven million members. Those members lobby and vote. Newspapers reach additional millions. Television and radio shows often pick up print stories and condense them into the most dramatic and simplistic messages. A little bad information goes a long way.

A well-crafted environmental bombshell can have sudden and costly results. The media stars of CBS's "60 Minutes" have often stirred citizens and governments to take immediate action against waste, fraud, unjust imprisonment, and a variety of other outrages. In February 1989, the show's Ed Bradley began the segment "A is for Apple" with a skull and crossbones on a beautiful apple, and announced, "The most potent cancer-causing agent in our food supply is a substance sprayed on apples to keep them on the trees longer and make them look better." With that simple statement, Bradley almost instantly delivered the bankruptcy papers to many apple farmers. He was referring to the chemical Alar. Not only is it deadly, he claimed, but those most at risk are . . . the children.

Some fifty million viewers saw the program. Mothers poured apple juice down the drain, threw apples in the trash, and started calling their congressmen and the Environmental Protection Agency. The next day, the Natural Resources Defense Council (NRDC) orchestrated press conferences in twelve cities. A week later, the actress Meryl Streep appeared on Phil Donahue's show in front of ten million viewers, mainly women, promoting her new group, Mothers and Others for Pesticides Limits. She urged the audience to buy a booklet on pesticides being sold by NRDC. Panicked viewers could also call a 900 number for more information. Book sales soared to almost 100,000 copies at $7.95 each. The 900 number earned Streep's group more than half a million dollars.

Few of the fifty million viewers could know that CBS had agreed to this promotion in return for exclusive first rights to NRDC's sensational report on Alar and cancer. Nor would anyone find out for almost six months that the deal between NRDC, CBS, and Streep had been set up by a slick political publicist named David Fenton.

Fenton has never been anything but an advocacy journalist. He began his career as a photographer for the Liberation News Service, edited the newspaper of the radical White Panther Party, and took on Ralph Nader, *Mother Jones*, and the Sierra Club as his first public relations clients. Whether working for American groups or the Marxist governments of Nicaragua and Angola, Fenton's specialty has always been getting radical views into the mainstream media. For a hot story, CBS went along.

In return for the sensational exclusive, CBS agreed to act as a fundraising front. To protect its benefactor, CBS excluded any scientific doubt about Alar, and there was more doubt than proof, despite the fact that Bradley called the NRDC study scientific and carefully researched. It wasn't. Independent scientists would later ridicule the study.

In addition, Bradley never mentioned that EPA's skull and crossbones were legally reserved for chemicals considered "dangerous" and that it had designated only its mildest warning for Alar. Nor did CBS tell its viewers that only 5 to 15 percent of U.S. apples were treated with Alar.

Of course, no one mentioned any benefits of Alar besides redder and riper fruit. The poisoning of our children was supposed to be the result of frivolity and greed. However, by keeping apples from dropping from the tree, Alar denies apple-eating bugs a place to breed on the ground. Fewer bugs mean fewer insecticides. One expert estimates that Alar reduces the use of other pesticides by 70 percent.[17] As a result, Alar becomes an important incentive for the pest-control technique that environmentalists value second only to organic gardening—IPM, or integrated pest management.

Even two years after the Alar scam had been completely unmasked, scientific fact made little difference to environmental advocates. Writers continued to talk about NRDC's victory over a dangerous pesticide. The University of California published *Toxics A to Z: A Guide to Everyday Pollution Hazards*. The four authors noted that "data on the health risks of toxics are scanty and laden with experimental uncertainty," but in their three pages on the dangers of Alar, their references list only an EPA fact sheet, a Sierra Club book, and NRDC's report.[18]

Despite the collapse of NRDC's scientific evidence, its media strategy worked. Fenton explained the report's objective in a memo:

Our goal was to create so many repetitions of NRDC's message that *propaganda tactics* average American consumers (not just the policy elite in Washington) could not avoid hearing it—from many different media outlets within a short period of time. The idea was for the "story" to achieve a life of its own, and continue for weeks and months to affect policy and consumer habits.[19]

Fenton had begun to plan the media event in October 1988. In addition to "60 Minutes," he had made agreements for interviews with *Family Circle*, *Women's Day*, and *Redbook*.

The goal of the Alar campaign clearly was not saving children from Alar, or even promoting greater use of toxicological research in protecting the public from dangerous chemicals. NRDC's callous disregard for objective science and its alliance with Fenton and his solidly Marxist view of America suggest that the real motive was to attack the whole idea of a private chemical industry.

The Alar fraud cost apple growers an estimated $200 million in 1989 sales. Many family farmers, apple processors, and packing cooperatives went bankrupt.

CONFLICTS OF ENVIRONMENTAL INTERESTS

The environmental movement expresses sympathy for ordinary citizens injured as innocent bystanders in environmental wars, yet it does not hesitate to run over a few innocents in its headlong pursuit of the bad guys, who are led by big corporate interests protecting their profits and fat salaries.

Undeniably, corporations and businesses lobby and fight to protect their wealth. Putting aside the issue of how many common citizens share in that wealth, we should pause to consider if the environmental movement's own advocacy journalism is immune from the same temptations.

Some 70,000 private and public companies with more than one million workers engage in some kind of environmental work. Their annual revenues for products and services have topped $130 billion. A list of a few of the Audubon Society's big donors for 1992 reveals a lot of money from corporations. Some may be looking for good will, innocence by association, or even some protection for their polluting activities. Many others make a nice profit from environmental regulations and therefore encourage the movement's activities. Consider this selection of Audubon funders:

$100,000 to $499,999
Brooklyn Union Gas Co.
Consolidated Edison Company of NY
Waste Management Inc.
Wheelabrator Technologies, Inc.

Over $50,000 in Gifts
Fuji Photo Film U.S.A. Inc.
General Electric Company
The Gitano Group, Inc.
IBM
Procter and Gamble
Turner Broadcasting System, Inc.

$10,000 to $49,999
Fred Stanback, Jr. [pharmaceuticals]
Barton Cotton Sales Corp.
Cargill Fertilizer, Inc.
Chevron USA Inc.
Conoco Inc.
E. I. Du Pont de Nemours and Co.
Hawaiian Electric Industries
J. P. Morgan & Co.
Parke-Davis, a Warner Lambert Company
Rockefeller Family Members
Shell Oil Company
Union Camp[20]

Should we believe that this kind of money does not influence the Audubon
Society, its lobbying, or its magazine? If not, why would such sums tempt
members of Congress or the business press? Either money does not tempt
them either, or Audubon people were born with a gene for saintliness. As
Daniel Koshland said one month after the "60 Minutes" Alar scam:

> It is time to recognize that public interest groups have conflicts of
> interest, just as do business groups. . . . Businesses prefer to be out of
> the limelight; public interest groups like to be in it. Because they are
> selling products in the marketplace, businesses downplay discussions of
> hazard. Because public interest groups acquire members by publicity,
> they emphasize hazards. Each group convinces itself that its worthy

goals justify oversimplification to an "ignorant" public. Businesses today have product liability and can incur legal damages if they place a dangerous product on the market. Public interest groups have no such constraints at the moment; it may be time to develop appropriate ones so that victims of irresponsible information have redress. Public interest groups, as well as apple growers, contribute importantly to our society, but both groups should be accountable for their acts.[21]

On the Audubon list, one company in particular stands out—Waste Management. Waste Management appears on the list in two donor categories: $100,000 to $499,999 and over $50,000. In the first category, it appears with its sister company, Wheelabrator Technologies. Waste Management runs landfills. Wheelabrator specializes in air pollution reduction and water filtration. They both belong to a holding company called WMX Technologies, whose president is Phillip Rooney. Rooney also appears in two donor categories: over $100,000 and over $25,000. At a minimum, Audubon seems to have received $375,000 from WMX companies and officers. Why?

Mr. Rooney and his companies may simply like the Audubon Society. However, the founder of WMX told *Forbes* magazine that every time environmental regulations get tougher, companies have to hire more experts to comply. Rooney told *Forbes*, "Regulation has been very, very good for business."[22]

CREATING THE EVENTS THAT CREATE THE NEWS

When the environmental movement cannot write its own news, it knows that dramatic action can dominate the media. A well-crafted image, along with quotable sound bites, often stays with the reader longer than the journalism that surrounds it. If the 1970s counterculture achieved real expertise in any field, it was the creation of the "media event."

The first big media events in the television age were staged by the civil rights movement. These sit-ins and marches were actual events in which victims of social injustices went face to face with their opponents at schools, swimming pools, restaurants, hotels, movie theaters, and parks. In contrast, most counterculture protests in the 1970s were symbolic or ritualistic media events.

The biggest media event of the environmental movement came in 1992, when it orchestrated the agenda for the star-studded Earth Summit in Rio de Janeiro. The summit was officially planned by the United Nations as the Conference on Environment and Development, but the meetings and background papers that determined the course of the summit were dominated by activists and politicians rather than scientists.

The goal was to create a conference of such great import that world leaders could not refuse to come. Interior Secretary Bruce Babbitt understood the process: "Gradually the process took on a life of its own, making it impossible for political leaders to ignore."[23] More than one hundred presidents, prime ministers, and other high officials came, legitimizing the politicization of science and guaranteeing enormous media exposure. Environmental activists made sure the media had a veritable circus of symbolic entertainments.

The assumptions of the summit mirrored those of the environmental movement. Global warming, along with the guilt of the rich, was accepted as the cause of the problems of the poor. The official document issued after the summit says, "The poor are often the victims of environmental stresses caused by actions of the rich."[24] Poor nations and rich environmentalists agreed that the first step toward a solution is for the industrialized countries to transfer a large portion of their wealth to poor nations.

The giant media event memorialized its success in *Agenda 21*, edited by the lawyer Daniel Sitarz, another of the huge volumes that the movement produces periodically as a comprehensive guide for public opinion. At the beginning of the modern environmental movement, the document was the Club of Rome's computer fantasy, *The Limits to Growth*, with its clumsy computer models projecting past trends into the future as if nothing could ever change. In 1980, the big shocker was Gerald Barney's *Global 2000 Report to the President*, which predicted worldwide catastrophe by the end of the century. President Carter made the report a centerpiece of environmental diplomacy. *Time* and *Newsweek* ran full-page articles. Four years later, a meticulous exposé of the report's cooked numbers, shaky sources, and bad science, *The Resourceful Earth*, edited by the economists Julian Simon and Herman Kahn, received little attention. In our society, messengers bearing bad news become heroes and heroines. Messengers bearing good news are ignored.

As environmentalists began their massive public relations campaign for the Rio summit, many scientists recognized that propaganda was again about to be passed off as science. They saw that committees meeting to prepare the summit's agenda had few scientists, and those who were included were seldom specialists in the area being discussed. The chemist

Michel Salomon helped organize a group of scientists to warn world leaders and opinion makers that the Rio summit would have noble motives, eloquent speeches, and a distinct antiscientific bias.

Salomon and other scientists staged an event of their own to demonstrate how scientists approach such problems as global warming, ozone thinning, and hazardous chemicals. In April, some fifty scientists from different disciplines gathered in Heidelberg to discuss the management of hazardous substances. Discussion inevitably turned to the upcoming Rio conference.

The scientists adopted the Heidelberg Appeal to Heads of States and Governments. The statement, eventually signed by more than eight hundred scientists, including sixty-two Nobel Prize winners, endorsed the summit's concern for the poor and its call for science to address social problems. However, the scientists warned "authorities in charge of our planet's destiny against decisions which are supported by pseudo-scientific arguments or false and non-relevant data."[25] The Heidelberg Appeal names no names, calls for no money, and predicts no catastrophes. Its signers were not asked by the press to comment on the Rio summit or to appear on the news and talk shows reporting on summit events and conclusions.

CHILDREN AND EDUCATION

If environmental groups talk about any one reason for saving the planet, that reason is children, the future generations. The long-term dimension of their advocacy journalism is advocacy education.

Advocacy education about the environment takes two paths. One is political vaccination, which assumes that children should be given truth, so that when they begin to learn how to ask questions and do scientific experiments, they will already know the answers. The second is "keep 'em primitive." In other words, don't let them lose those childhood fantasies. Encourage them to talk to trees and to think of animals as their equals who just never learned to speak human language.

The political vaccination takes the form of a catechism. While the environmental movement is too sophisticated to require rote memorization of something like "two legs bad, four legs good" (the slogan of the ruling pigs in Orwell's *Animal Farm*), several "truths" appear over and over in children's lessons, television, and books. They include:

1) Recycling is always good.
2) Forests are disappearing, development is the bad guy.
3) Global warming threatens life on earth.
4) Greed causes environmental problems.
5) Nature is always right.
6) Primitive people were smarter managers of nature than we are.

Putting aside the question of truth, I wonder how kind or smart it is to scare the hell out of little kids. Four of these six messages are scary. Even if all of these messages were true, telling children about dangers they cannot avoid is a form of child abuse. At times, we do not even tell adult hospital patients all the bad news about their illness.

Even religious people who threaten children with burning in hell are at least talking about another world, one for which they have not yet booked a ticket. We may teach children the Ten Commandments and expect them to accept those rules before they can fully understand the reasons, but at least a child can intuitively understand such maxims as "Thou shall not murder" and "Thou shall honor thy father and mother." Children know pain hurts. And they understand that someone besides themselves must run the household. Our society has no serious quarrel with the Ten Commandments. But the commandments and truths of environmental education are subject to a lot of qualifications. Consider how some of them are taught and what the facts are.

RECYCLING

"Thou shall recycle" could stand a lot of important qualifiers that young children cannot understand and older children seldom get the chance to consider. From teachers' guides to EPA's education program, students of all ages are taught recycling as a commandment, not as an option for solving environmental problems. In fact, recycling frequently costs more than it saves, and while it usually preserves resources, it sometimes wastes them. For example, the use of reusable cloth diapers may save trees, but it also wastes energy through both production and repeated washing. Further, cleaning requires lots of water and detergents. The local sewage plant must increase its capacity and chemical use. Recycling paper also saves trees, but the chemicals necessary to remove ink and stains create more pollutants than the chemicals used to process new wood.

Both children and adults hear over and over again that we are running out of landfill space. The popular children's book *Fifty Simple Things Kids*

Can Do to Save the Earth has sold almost a million copies and claims that "we are making so much garbage that in many places there is not enough room to bury it all."[26] But most garbage does not pollute. It consists of paper, yard clippings, stumps, dirt, concrete, tin, aluminum, and plastic. We are not running out of space at all. The problem is more economic and psychological. Who wants to live near a dump? And how much will we have to pay to haul garbage to a dump site? *Landfill areas being used for housing.*

Environmental education does not teach thinking; rather, it teaches obedience to unelected, uncriticized authority. This is the kind of education that should be reserved for basic religious and social values such as the Ten Commandments and civil rights, concepts about which there can be little doubt. John Javna, the author of *Fifty Simple Things,* estimates that children send him 10,000 letters a year. Javna told the *Wall Street Journal* that "he detects an almost religious fervor."[27]

FORESTS ARE DISAPPEARING

Popular books and movies such as *Fern Gully* tell kids that the bad guys are aiming to clear every last tree from the face of the earth and there are only a few left. The message about the spotted owl forests of the Northwest is that it is now or never to save the last old-growth rain forests north of the tropics. Everywhere else developers are also paving over paradise and forests are vanishing. Almost all of these claims are highly exaggerated or simply false.

About redwoods and old growth, we hear only that the last 10 percent is about to be cut. The last 10 percent of what? Untouchable national parks contain millions of acres, including 109,000 in Redwood, 403,000 in Sequoia, 2.2 million in Yellowstone, almost a million in Olympic, and 235,000 in Mount Rainier. Washington State claims 2.5 million acres of public park, forest, and other land locked away as officially protected wilderness. Oregon has 2 million. The proper questions about old growth are how much should we preserve, where, and why?

From Pennsylvania to Maine, we now have more forests than we had fifty years ago, and the tree cover is increasing. A large part of the reason for this increase is that we have replaced the horse on farms and roads with tractors, autos, buses, and trucks. Every horse and mule used for power required some two to three acres of cleared pasture land. Even in Europe, where children are told that air pollution has created "forest death" or *waldsterben,* scientists have found what journalists and textbook writers do not want to hear—forests are expanding. A Finnish study of Europe's trees found that between 1971 and 1990, growth increased by 30 percent and would probably continue to increase.

The point is not to say pollution is harmless. It could already be weakening forests. The point is that in the real world, nature is a very complex and interesting system. In the world of eco-education, nature is simple and human beings are simple-minded saints and devils. We are not educating children to solve real problems.

PLASTICS ARE BAD FOR THE ENVIRONMENT

When I was working with my Episcopal diocese on environmental issues, I asked why everyone was so dead set against Styrofoam cups. The ready answer: because they are plastic, made with harmful chemicals, and use up landfill space. *Fifty Simple Things* encourages kids to "stamp out Styrofoam" because "using Styrofoam means using up precious resources . . . and adding more garbage to our world."[28]

What educators should be doing is teaching students how to analyze choices between plastics and, say, paper. That's science, not propaganda. When *Science* reported such an analysis, the chemist Martin Hocking considered the energy used in production, the landscape impact of getting the raw materials, air pollution, recycling costs, and greenhouse gases. Among other things, he found the following:

- Foam cups require a sixth of the amount of nonrecycled chemicals in production.
- "The paper cup consumes about 12 times as much steam, 36 times as much electricity, and twice as much cooling water."
- The greenhouse gas effect of a foam cup's pentane is less than that of the methane released by a landfilled paper cup.[29]

A more appropriate message to children about this subject might be that substituting synthetics for natural products is a strategy that sometimes saves more than it destroys. We are fortunate to have choices.

PESTICIDES ARE UNNECESSARY AND DEADLY

Many chemicals used to kill bugs, fungi, and weeds can also kill or harm people, especially in large doses. If one day we can grow crops without them, it will be because we have genetically engineered plants to produce their own internal pesticides.

Although organic gardening has contributed to a reduction in the use of pesticides, not a single study shows that its methods could produce food to feed the world. Children need to know that in large quantities pesticides are poisonous, but that they are necessary in safe quantities. They also need to understand that plants produce natural pesticides, many of which cause cancer in high doses. The choice is not between a chemical-free organic world and a toxic world. The choice is between the skillful use of chemicals and costly fear.

One of the older examples of this nonsense is that cedar wood is nature's safe way to prevent moths from eating wool clothes. First, cedar does not prevent moths from eating clothes unless the cedar makes a box or closet that can be tightly sealed against the entry of moths. Metal would do just as well, although perhaps moth larvae already in the clothes are weakened by the cedar aroma. Second, as far as I know, the aromatic chemicals in cedar are no more kindly to humans than the compounds in moth balls. Red cedar smells nice, but whenever I cut it or sand it, I am seized by violent sneezing, stuffed sinuses, and running eyes. Do I really want to spend one-third of each day sleeping in a room where the closet gives off a constant low-level stream of such powerful chemicals?

Corporations Destroy, Environmentalists Save

The greatest purveyor of this idea is the movies, and its most successful practitioner for children is Steven Spielberg, the director who never grew up and accepted the complexity of the world in which children must eventually live.

In *Jurassic Park,* the message is not that people can be corrupted, but that science and industry are corrupt by definition: science, because it has no heart, corporations, because their purpose is to make money. In the movie, Spielberg turns scientists against their own profession, a most persuasive trick. The paleobotanist played by Laura Dern tells the park's designer, "You can't control nature!" The hip mathematician spouting one-liners about chaos theory says that science is "a violent, penetrative act." In other words, scientists rape nature just like other bad guys.

Carrying on in the tradition of Rousseau, Wordsworth, and Thoreau, environmental movie makers have paired the industry's early sentimentalism about nature (à la *Bambi*) with stereotyped symbols of evil—scientists, businessmen, and lawyers.

Children must be taught to recognize evil, of course. Great religious literature has done this quite well. The difference between Hollywood environmentalism and the Koran, the Talmud, or the Bible is that the sacred books do not teach that particular professions (other than prostitution and thievery) are evil. A rich man may have great difficulty entering heaven, but not because he is inherently evil. Discussing the power of the movie industry, the English professor Carol Muske Dukes said:

> Whether the next generation uses Big Science to our ultimate benefit or detriment may well come down to one question: Which applied science has a better chance of manipulating our images of the future, gene-splicing or big-screen special effects? Think fast. You have exactly three minutes before the planet blows up.[30]

It is hard to judge the consequences of simple-minded environmentalism on education and what students will do in a few years when they become voters, governors, legislators, and presidents. Lenin is supposed to have said, "Give us the child for eight years and it will be a Bolshevik forever." But perhaps he was no wiser about children than about the wonders of communism. When reality eventually erodes the last supports of false idols, they often collapse in harmless dust.

It is one thing to present recycling and organically grown foods as interesting new choices, and something else entirely to present them as proven solutions that eight-year-olds must accept on faith. Most environmental educators with wide influence leave no doubt that they are promoting "truths" rather than curiosity or science.

Mike Weilbacher, a Philadelphia news broadcaster, rejoices that the annual Earth Day sets the tone for America's environmental education: "It is a day of extremes, for mourning and dancing, for educating and advocating, for dressing as a timber wolf and lecturing on global warming. . . . *But mostly, it is a day for kids* [emphasis added]."[31] Weilbacher calls Earth Day "the new children's crusade."

Environmentalism has not yet come under the ban against teaching religion in school, but maybe it should. Earth Day is a good idea, but it has been dedicated to propaganda rather than thinking. It has done what its opponents consistently have failed to do—interest the American public in environmental issues. Naturally, its creators have taken possession of its contents. The former Wisconsin senator Gaylord Nelson, who helped invent Earth Day, says, "Every kid in 12th grade has now been through some kind of Earth Day experience, and they're getting formalized education in increasing numbers."

The young Harvard student who organized the first Earth Day for Nelson was Dennis Hayes, who continues to take part in it. Hayes notes that the environmental movement pays a lot of attention to charismatic animals such as bears, wolves, pandas, and whales, "but there is no more charismatic megafauna than your eight-year-old daughter. Hell, there's nothing someone my age wants more than to be a hero to your child."

Like the consumer product advertisers who environmentalists despise, they themselves know that the way to sell ideas and products is through children. Hayes says, "Watch out for these kids; they may be the fourth wave, washing in on the rising tide of environmentalism, pulled by the gravity of Earth Day."

CHAPTER 7

Nature as We Want It: Can the Environmental Movement Adapt to the New Ecology?

But when we defeat an enemy in battle—when we defeated Germany and Japan in World War II—do we simply go on killing and slaughtering? Of course not. We have defeated nature. We must do as we do with a defeated nation—help nature, and recognize that we must live with nature from now on, forever. The war is over.

—Eugene Odum, Professor of Ecology, University of Georgia

When the environmental movement began searching for its identity in the 1970s, its members often called themselves ecologists. The public went along, sometimes calling the more flamboyant members ecofreaks. Newspaper reporters wrote about the "ecology movement." Professional ecologists—that is, working scientists—did not protest in any audible fashion, but they should have. Being able to distinguish ecologists from environmentalists is as important as knowing the difference between surgeons and witch doctors.

Ecology is a science. It produces knowledge. Environmentalism is a political and religious movement. It produces judgments and laws. How much the movement helps the world will depend on how well it understands science and how wisely and honestly it uses the knowledge that science offers.

The historical value of any movement sooner or later depends on how well it can absorb new knowledge—especially knowledge that challenges its basic assumptions. History is littered with the remains of governments, social systems, and religions that could not or would not absorb new knowledge, or even read the handwriting on the wall.

Most environmentalists have college degrees and think of themselves as sophisticated and open-minded. Like all true believers, however, they have assumptions that limit their ability to absorb new ideas, assumptions that define their friends and enemies and by which they know what is right and wrong. Those assumptions are as easy to summarize as the Ten Commandments. Like the commandments, they are often bent or ignored to suit personal convenience or practical purposes. Nevertheless, they are the principles that the environmental movement wants written into law and regulation:

1) Nature is good.
2) Altering or destroying any part of nature is bad.
3) Nature has a balance that humans always disrupt.
4) The more power humans get, the more damage they do to nature.

The test of whether this summary is fair is to read popular environmental magazines and books and ask if anything in them contradicts these statements. The vice president's book, *Earth in the Balance*, although fitted with politically prudent escape hatches, more than once pledges allegiance to each assumption. In environmental books, videos, films, and magazines, exceptions and dissent are as rare as whales in the Hudson River. Affirmation is everywhere.

Environmentalists are almost always political liberals who have little sympathy for Christian fundamentalists. Even Christian environmentalists often buy the idea that Christianity has sinned against nature and needs to be more like primitive animism or oriental mysticism. Yet environmental faith is quite rigid and its story, biblical. History and human nature as described by environmentalists match the story of the Garden of Eden and the sinfulness of Adam and Eve and their descendants. Innocent preindustrial, prescientific cultures show what we could have been, while civilized cultures show how low we have fallen. The fruit that Adam and Eve ate gave them knowledge, and knowledge divided their world into good and evil. By the time Noah appeared, "the earth was corrupt in God's sight and the earth was filled with violence." And so it is in the environmental vision: knowledge begets evil and violence.

In Eden and in the environmental story, the harmony of nature and its ability to nurture humankind are destroyed by ambition, greed, and sin. These sins are inflicted on nature and native peoples almost exclusively by white males. Political correctness has absolved Eve of original sin and endowed women with a special sensitivity to nature.

Vice President Al Gore seems to have converted from Southern Baptist puritanism to environmental puritanism. He too blames white male dominance, and his "dysfunctional family" is not unlike Adam and Eve's family, which not only corrupted the earth but in which Cain also slew his brother, Abel. Gore talks about a time among early people when there was a "reverence for the sacredness of the earth" and a "harmony among all living things." Nature by itself, he writes, runs on "principles of balance and holism." Nature produces toxic substances, but it maintains a "balanced and mutually beneficial relationship between the species involved." He says that the environmental movement is "engaged in an epic battle to right the balance of our earth." Then he redefines ecology: "Ecology is the study of balance."[1] He declares that we should apply some of these principles of balance to our political system.

The vice president is not alone among politicians. In July 1989 in Paris, leaders of the world's seven largest industrialized democracies met to discuss the global environment. Their meeting ended with a call that could have been issued by any environmental group: to protect nature's fragile balance.

The idea of a balanced nature makes all the other basic environmental beliefs logical. But suppose no balance ever existed in nature? If nature is not balanced, then a fundamental assumption of the environmental movement, and all that stands on it, becomes illogical. And dangerous. If the conditions of nature fluctuate wildly and seldom repeat, then human attempts to stabilize it are egocentric and selfish. Working on the balance model, environmentalists ask us to spend our time and money to sustain a world that does not exist except in environmental scripture.

So much that the environmental movement asks for is in the name of the "balance of nature." Al Gore writes that our greatest problems are caused by a "new relationship between human civilization and the earth's natural balance."[2] The environmental movement's attack on Western civilization assumes it has upset some ideal steady state. A 1992 meeting of scientists and religious leaders in Washington, representing some 330,000 congregations, declared we must maintain the earth as we received it. From the vice president's office to Podunk, environmentalists are staking their fame, fortunes, and sacred honor on something that does not exist.

Environmentalists talk about needing a new model of nature, but they certainly do not want the one emerging from today's ecology. The ecologist Dan Botkin, a small man with a gentle manner, delivers the message of modern ecology with quiet patience, something like a UPS man delivering Pandora's Box. Botkin says that nature's preference seems to be change— often unpredictable and radical. There is no balance. There's no steady state, not even a steady seesaw state with a constant center. Whole populations of plants and animals flourish and then crash with a violence that would appall most environmentalists. Nature has no preference for life as we know it, or life in any form. The landscapes we visit and paint are short-lived in natural history. The only way to maintain a balance in nature is through sophisticated, often high-tech intervention, such as managing the flow of the Mississippi River or building earthquake-proof skyscrapers.

Botkin poses a simple question that is the biggest challenge to conservation thinking in our time: "How do you manage something that is always changing?"[3] The message of modern ecology for environmentalists is action: if you do nothing, you will get something you do not expect. In the words of Nietzsche:

> You want to LIVE "according to nature"? O you noble Stoics, what fraudulent words! Think of a being such as nature is, prodigal beyond measure, indifferent beyond measure, without aims or intentions, without mercy or justice, at once fruitful and barren and uncertain; think of indifference itself as a power—how COULD you live according to such indifference? To live—is that not precisely wanting to be other than this nature?[4]

SCIENCE CHALLENGES HISTORY

Like environmentalists, Botkin talks about spaceship Earth. But there is no automatic pilot. And God is not our copilot. We are on our own and we have to learn how to fly. Even before we chart the course, we have to invent instruments to monitor systems and install controls for steering. It is as if among all the world's life forms, the question suddenly arose, "Who's in charge here?" And the human animal answered, "I am."

For many environmentalists, such a suggestion verges on blasphemy. Blasphemy insults the sacred, and the idea of balance in nature has been a concept almost as universal as God. Thus, the New York–based Natural Resources Defense Council reprints an article from the *Toronto Globe* in which a Cree hunter says, "The earth was created the way it was by the Creator and changing it is unnatural and wrong."[5] The idea of a balanced creation repeats itself a thousand ways in popular culture.

According to environmentalists, nature's balance has been disturbed by civilization, but if nature could regain control, beautiful order would reign again. They have faith that nature unshackled can be counted on to restore its own balance. They do not believe that humankind can lead the way to a better environment. As Botkin has experienced:

> There are groups that are really anti-science, anti-knowledge because they think that's what's messed things up. I was interviewed by a reporter the other day who said, "You seem to have a bias toward scientific knowledge."[6]

The balanced nature myth has special appeal to people who no longer have traditional religious principles to explain what is right and wrong. Nature takes the place of God. It is no accident that American churches embraced the environmental movement only in the late 1980s. The more the movement dropped science in favor of "eco-spirituality," the easier it became for liberal churches to identify with it.

Along the way, the multicultural movement made it possible to accept primitive religions with their animal gods and their conversion of geology into theology. The Catholic priest Thomas Berry, the most popular churchman among environmentalists, has said that Christians must "put the Bible on the shelf" for a few decades while they develop a new sense of God and nature.[7] As if to declare that nature belongs to God, not humankind, a mainstream anthology of environmental thinking begins with this quotation of God's words from the Book of Job: "Where were you when I laid the foundation of the earth?" With the churches joining the movement, the balance of nature truly became a sacred principle.

The balance of nature satisfies our wishful thinking and longing for more order and permanence than our tumultuous human history offers. The sense of a sacred nature offers a belief that is immune to the refinements and reversals that scientific progress demands. However, most ecologists now believe that such a balance has never existed. "The idea makes good poetry but bad science," said one ecologist. At Ohio State University Peter Chesson, a theoretical ecologist, says that the idea of nature having an equilibrium is "dead for most people in the scientific community." The University of Minnesota's David Tilman says that among ecologists, *balance of nature* is a term that "hasn't been used much in twenty years."[8] For ecologists, the orderly and harmonious world in which wolves and deer, hawks and rabbits, bees and flowers keep each other's populations in balance has gone to the old-age home of quaint ideas to rest alongside spontaneous generation and astrology.

THE "HANDS OFF" ILLUSION

In many cases, "hands off" environmentalists have been playing God as surely as former vice president Dan Quayle's God Squad, the giant timber companies, or the government managers whom they increasingly distrust. To demonstrate the unexpected results of letting nature "manage itself," Botkin sometimes uses a picture of Rutgers University's virgin Hutchinson Memorial Forest, a remnant of mid-Atlantic oak-hickory forests. Rutgers purchased the forest in 1954 with great fanfare. *Audubon* and *Life* magazines celebrated with beautiful pictures. In the seventeenth century, a Dutch naturalist said that one could easily drive a horse and wagon between the large trees in such forests. The forest in Botkin's recent picture has few large trees, though, and even a trail-bike rider could not penetrate the thick undergrowth and vines. Letting nature manage the preserve has meant allowing hurricanes to work their will and standing by while alien species such as Norway maple and honeysuckle invaded. Where Native Americans once hunted and cleared brush by burning, fire has become rare.

At the heart of the great debate over the Northwest's old-growth fir forests lies a romantic notion of their eternal presence. However, if nature had its way, these fir forests would be gone. Without the millions of dollars spent on smoke jumpers and fire-fighting programs, most of them would have burned long ago. A few would have survived to become climax forests (which are not Douglas fir forests). Some estimates of fire events in the Cascade range from California to Canada calculate that nature's fires burn over the entire area during the course of some six hundred years. The fires, especially in artificially protected stands, are often devastating to wildlife and nearby streams.

Old-growth forests often do not match the pictures in coffee-table environmental books or the most popular national park sites. Douglas firs grow naturally in uniform stands. Mature stands often have massive blow-downs, trees toppling each other, dead trees bringing down live trees and crushing young trees. A blow-down often looks worse than a clear-cut.

REDEFINING NATURE

Even when we choose to manage nature to maintain a "natural state," Botkin and other ecologists say that the popular definition of nature has little to do with the real world. Botkin says that if we were to ask what the natural state of the great Boundary Waters wilderness is, he "could argue that its natural state is pure ice."[9]

The ancient balance of nature that many people love and want to save is often a figment of their imagination. In the 1992 movie *The Last of the Mohicans,* the action takes place in what viewers thought was a primeval wilderness where the Indians had lived for thousands of years without disturbing the great forest. The movie opens and continues to the end almost entirely in deepest woodland shade. Most viewers thought this was virgin forest the way colonists found it. This primeval forest, however, was not in the Mohican's New York but in North Carolina. The virgin forest had not been a virgin for one hundred years. Most of the trees averaged some twenty years old.

Visitors to St. John National Park in the U.S. Virgin Islands fall in love with its green hills and wildlife. Few recognize that these tropical forests were once clear-cut by Danish sugar planters and that what they see is not some untouched piece of paradise. Park Superintendent Marc Koenings once had friends come to visit who he knew would be interested in the natural history he was working so hard to restore. He was sorely disappointed when the first thing they wanted to see was the wild donkeys. The St. John donkeys are the descendants of imported plantation donkeys, and they have devastated the island's native vegetation. The second greatest tourist wildlife attraction is the mongooses, which are more plentiful than people. They too were introduced by colonists. They were supposed to control the rat population, but because rats are mainly nocturnal and mongooses work during the day, the twain seldom meet. The mongooses have found local bird nests to be a ready source of protein.

What most environmentalists want to preserve is not pristine or productive nature, but picture-book nature. Picture-book nature is what pleases our eyes and our fictional sense of what is valuable. For environmentalists, even the harshest deserts and the polar regions are likable fantasies, a kind of science-fiction landscape. Environmentalists have gutted and stuffed nature in the same fashion that they have turned primitive people into wooden Indians.

Where I live, near North Carolina's booming Research Triangle (three major universities with a six-thousand-acre commercial research park between them), the rural South is being resettled by university-educated people from all over the world, but mainly from the urban North. The environmental change that most riles them is clear-cutting. I have been called at least twice a year by someone who wants to stop the logging being done on someone else's land. Their reason to stop the logging is to save nature.

The nature they want to save is generally a growth of pine trees twenty to forty years old. Almost 100 percent of these pine stands have grown on old fields. Many were planted like crops. These yellow pines are nature's weeds, opportunists and squatters. The natural old-growth forests here in the rolling piedmont hills feature beech, oak, and hickory.

I do not like the way clear-cut lands look either. My neighbor recently wiped out several hundred acres including two-hundred-year-old beech trees that were worth more for their real-estate appeal than for the pulpwood they made. But these are matters of scenery and economics, not science and nature. The pine woods do have a certain hush to their needle-carpeted floor, a deep, soothing shade, and a nice music in the wind. But they do not have much wildlife. I collect wild mushrooms, and these pine stands are almost always barren. Even the mature hardwood areas do not support a great variety of wildlife. Any bird feeder in town or open farmyard has ten or twenty species visiting, while mine in the woods has only four or five. To add insult to injury, clear-cut lands in this area support more actual biomass and take more carbon dioxide out of the air faster than a mature forest. There are not as many turkeys or bees, but mice, voles, snakes, rabbits, foxes, hawks, songbirds, and blackberries live in greater abundance in clear-cuts. My neighbor has impoverished the landscape, but not nature. And I dare say that any good artist zooming in on the new growth would find plenty of vitality and interesting shapes to paint.

As for the forest, it will return quickly. I don't enjoy looking at the dense thickets that spring from clear-cuts and dominate for the first ten years, but I know life is thriving in there. A pine stand will come up with pine. A hardwood forest will regenerate from the stumps faster than it can grow from plantings. In the overall picture of American forestry, the total volume and cover of trees continue to increase along with many forms of wildlife.

The failure to understand nature's true personality and its unpredictable cycles has caused environmental "crises" for a lot longer than the modern environmental movement has been around. In the late 1860s, New England fishermen rose up in protest against net fishermen. They deluged the legislatures of Connecticut, Massachusetts, and Rhode Island with petitions. More than 11,000 people signed petitions in Massachusetts in 1870. The head of the U.S. Fish Commission concluded that netters were about to wipe out entire species. Despite the fact that the netting continued, the species came back in abundance. The protests had begun during a natural crash in fish populations. They faded out with the new population boom.

This tale should not be taken as a blithe "what, me worry?" approach to resource consumption, but it does suggest that regulations should be more closely synchronized with and adaptable to nature's own squiggly course through time.

Botkin says that the more his views about nature began to diverge from tradition, the more he wanted to know how tradition shapes social and scientific expectations. Why was it, he asked himself, that when his colleagues were asked to formulate public policy, they often gave "advice contradictory to the evidence they knew?"[10] In 1978, the Woodrow Wilson Foundation paid him to spend a year "to read what everybody from the Greeks to the 18th century said about nature." That year in the humanities helped Botkin understand the growing gap between modern ecology and society's assumptions about how nature works.

WILDERNESS LOST

An early casualty of the new ecology was the idea that wilderness is a model of nature's ways. The supposed virgin ecosystems of the past have been the ideals that have driven environmental activism. "In wildness is the preservation of the world," Thoreau proclaimed. Alas, the wildness that Thoreau saw in New England may not have been wildness at all. New evidence from historians, paleontologists, archeologists, and geneticists has clobbered the notion that Western materialism destroyed a natural paradise. Much of the world's most admired "wilderness" may actually have been the intentional or de facto creation of humankind. The great herds and predator populations of Africa's Serengeti savannahs, for instance, may owe their abundance to the primitive practice of using fire to maintain grassy fields for hunting. Amazonia's "virgin" rain forests have been extensively populated by humans for more than ten thousand years, and these humans imported, planted, and transplanted a variety of plants, and perhaps introduced new species of animals.

Lee Talbot of the World Resources Institute has worked in more than one hundred countries and has come to the conclusion that "in a real sense, human beings have been changing the face of the Earth since their earliest times."[11] The Yale historian William Cronon says that Native Americans exerted a huge influence on the forests. Nature itself, Cronon says, brought "environmental changes on an enormous scale, many of them wholly apart from human influence." He concludes, "There has been no timeless wilderness in a state of perfect changelessness, no climax forest in permanent stasis."[12]

Scientists who see nature from a Third World perspective have welcomed the recognition that humans have always played a role in nature as we know it. To them, the ecology of change liberates environmental thinking from the elite urban fantasies of northern industrialized environmentalists. In a recent issue of *BioScience*, the botanist Arturo Gomez-Pompa and the anthropologist Andrea Kaus argued that the intense interest in climax forests and their value as wilderness preserves "represent mostly urban beliefs and aspirations. All too often they do not correspond with scientific findings or first-hand experience of how the world works."[13]

MAKING CHAOS OUT OF ORDER

The complexity and speed of change even when humankind is not playing with bulldozers, chemicals, or fire are illustrated in the story of a simple weed. Recording the behavior of several plots of the midwestern "pant creeper" (*Agrostis scaber*), two University of Minnesota scientists showed that even over a period of a few years, the biosphere, like the stars of the universe, might tend more toward chaos than our present sense of order. In 1985, David Tilman and David Wedin sowed pant creeper in a variety of soils. By 1988, the pant creeper population in the most fertile plot showed a six-thousand-fold explosion; then it crashed to near zero. Other plots also showed a variety of unexpected results.[14] Botkin says that twenty years ago these results would have disappointed researchers and been dismissed as a failed experiment. Since then, a concept developed in physics and astronomy has allowed us to make sense, if not order, of such results.

Chaos theory holds that any small differences in the initial conditions of a system will tend to be magnified over time. For example, two grass systems beginning in soils with small differences in fertility may look alike at first, but eventually those small differences will make the two systems quite different from one another. We can see evidence of this principle in bacterial colonies, coral reefs, and American cities.

According to chaos theory, natural systems do not behave like a well-ordered machine. A galaxy in space, or even grass in a few cubic meters of soil, undergoes huge changes that seldom if ever repeat in an orderly pattern. There may be an underlying long-term pattern, but it is not deter-

mined by the attraction of some immovable point of balance. Neither is it
entirely random. If it is not entirely random, perhaps it is ultimately predictable, but not with our current tools.

Tilman's surprise at his findings is a scientist's miniature of the conversion that must happen among environmentalists and society at large. "I
never imagined I would find chaos," he told the *New York Times*. "I imagined it would grow up to equilibrium. This has changed my world view, to
be blunt about it."[15] Further, the more species an ecosystem holds, he says,
the more likely we are to find that instead of balance there is chaos.

Chaos theory has become a thorn in the side of environmentalist
philosophers as they attempt to find new causes to rally the public. Environmentalists do not deny that nature changes, but it does so on the slow,
predictable scale of evolution and geological movement. In a 1992 *New
York Times Magazine* story about the worldwide decline in some frog populations, the author quotes the wildlife biologist James Vial of Oregon State
University. Vial tells millions of Sunday readers, "Something is out of balance. The changes in the environment may be more drastic than they
appear."[16] While most herpetologists see the decline as a global disaster, a
few suspect it might be a normal part of nature's chaotic behavior. In the
same article, Peter Morin, a community ecologist at Rutgers University
says that science does not know if the present decline is really worldwide,
or whether frog populations are decreasing permanently or just going
through a large but natural fluctuation.

The sudden increase in graduate students studying frogs also accounts
for the increased number of reports. The students begin their studies when
they find a population. A few years later if that population goes into a natural decline, the observers report a problem. Frogs are often plentiful and
their lives short, so they make good subjects for the study of population
dynamics. That field became increasingly popular as ecology attracted
more students. Scientists who began studying frog populations since Earth
Day 1970 are quite likely to be discovering now that frog populations fluctuate suddenly.

How widely do frog and salamander populations fluctuate? Widely
enough for fluctuations to look like disasters—or miracles. A study of three
salamander and one frog species at the Savannah River Ecology Laboratory in South Carolina found that between 1978 and 1990 a salamander
that was almost impossible to find at the beginning of the study multiplied
to six hundred breeding females by 1990. Other species also increased or
held steady.

The scientist who wrote up the Savannah River study for *Science* said that his unusually good breeding area does not prove that pollution is not killing amphibians in other parts of the world, but, he told the *New York Times,* "my money is on the natural fluctuations."[17]

Several reasons explain why we have not seen the real dynamics of change in biological systems and why the ideal of balance still guides public opinion. William Schaffer, a pioneer in the application of chaos theory to biological systems, says that the signs are much harder to detect in a forest or ocean than among the stars and galaxies. Rather than dealing with the relatively neat measurements of energy, gravity, and motion, the variables in an ecosystem are more numerous and more difficult to observe and measure.

THE CHALLENGE TO MANAGEMENT

The environmental community is hesitant to catch up with ecology. After Botkin argued his case in a keynote address at the American Forests' 1992 forum, "People as Positive Agents of Environmental Change," the panelists who followed went on to talk about preserving the balance of nature and not letting science and technology obscure spiritual values. While Botkin argued that resource decisions have to be opened to users such as loggers, fishermen, and hunters, other panelists talked disparagingly of "Joe and Jane Sixpack," of converting the unconverted, of getting industry's attention by hitting it with a figurative two-by-four, and of making leaps of faith. They left the clear impression that for them, nature is a set of beautiful landscapes and objects, not processes. The dominant theme of "People as Positive Agents" was keeping people away from nature and its management. Despite Botkin's cogent argument for sophisticated studies of environmental changes using satellites and computer analyses, most panelists still suspected that computers are a trap, a way of preventing intimacy with nature. Yet computers are now one of the most basic tools for understanding how nature works. If we recognize the complexity of the environment, how can we suspect computers of tainting our understanding? Some computer operators, perhaps, but not the technology.

For the new ecologists, the realization that nature has no balance means that managing the biosphere is much more complicated than we ever imagined. To keep from making ever greater mistakes as human populations and powers explode, Botkin says, "we can no longer rely on nineteenth-century models of analysis for twenty-first century problems."[18]

Clearly we cannot rely on the Romantic fictions of a kind and unchanging nature in which the lion and the lamb would sleep side by side were it not for the corruption brought on by humankind's knowledge and greed. Honest science runs into enough obstacles as it is. If environmentalists are really in a hurry to solve environmental problems, they must take off the rose-colored glasses and sit down at the microscope, telescope, and computer.

Botkin and others say that while we may have the wisdom to act prudently, we are far from having the facts for solutions. Many claims passionately presented as fact are actually emotional biases or intuitive guesses. In the Pacific Northwest, for instance, environmentalists assert that forests are being lost forever, but no one has reliable data on how much forest is regenerating, which would provide evidence to determine what should be done. Botkin recently started to study the relationship between forests and salmon in Oregon. Despite claims that timber cutting is destroying the five species of salmon that frequent the rivers, Botkin says that no one has maps of forest regrowth and no one has been tagging natural salmon populations. Little is known of the life cycles of natural populations. Environmentalists assumed a balance had been destroyed. Industry assumed a balance would reestablish itself. No one assumed that hard facts might help.

Oddly enough, while scientists have been methodically recording changes in the heavens for centuries, no one has been making the same careful observations of forests. Our inability to say how much forest is growing in the Pacific Northwest is a small corner of our ignorance about forests. With everyone talking about sustainable development, it would seem that at least local information on forests could be found. Botkin says that for twenty years he tried to get a reliable set of statistics on sustainable forest use. He looked for a forest that had had three harvests, the third being equal to or greater than the first. Over and over, he was assured that forester Jones or Smith over in such and such a forest had the data. But each time he followed a lead, the data turned out to be inconclusive or nonexistent. As a practical matter, Botkin says, there was "no basis on which to have a discussion."[19]

HOW TO ACT AMIDST UNCERTAINTY

When we ask what should guide our intervention in the natural world, our dilemma is this:

1) If we are to choose only what nature would choose, how can we choose opposite paths?
2) If we choose one path, which principle will tell us which path that is?
3) If we make our own choice, what moral and practical principles should guide us?

For more than four billion years, nature has made a flowing kaleidoscope of choices. Religious people from the Amazon to Riverside Church might demand that we leave the choices to God. But whose god? Even fundamentalists cannot agree on what God says or has said. Some people say God has told them to go to war, others say God has told them to be pacifists.

No matter what the religion, though, God has chosen to set human imagination loose in the world. Perhaps God has created nature as the context for civilization. Perhaps God has also set in motion the process that led to civilization—agriculture, urbanization, the Industrial Revolution. The human mind is a product of nature. Although we have thought of many means of improving it, and we try to optimize its usefulness and pleasures, our brains are one of the few natural creations that we have not been able to change very much. We have, however, learned an enormous amount about how to extend its powers.

Accepting the fact that nature constantly changes, often in unpredictable and undesirable ways, does not mean that we have to accept all change. Botkin says, "We must focus our attention on the rates at which changes occur, understanding that certain rates of change are natural, desirable, and acceptable, while others are not. As long as we refuse to admit that any change is natural, we cannot make this distinction and deal with its implications."[20]

Looking for environmental solutions without understanding the kaleidoscope of changes is a futile task. It is the situation Botkin found at the beginning of his study of salmon and forests in Oregon. Trying to manage resources like these without adequate data is "like suddenly being thrown into the cockpit of an airplane with no instruments," Botkin says. Public policy, he declares, is being made on pitifully weak data.[21]

To expand efforts to apply ecology's insights to environmental problems, Botkin founded the nonprofit Center for the Study of the Environment in 1992. The center's guiding principle is that human society is a part of nature: "Humans are an integral part of the ecology of the planet. The only lasting environmental solutions are those that take into account the dynamics of human society as well as those of natural systems."[22] At the core of the center's methods are computer projections that incorporate data from satellites as well as field studies and that are then used to model the likely consequences of political and business decisions.

This approach worked for Botkin in 1987 when he convened a blue-ribbon panel of scientists to study the impact of Los Angeles's huge water withdrawals from Mono Lake in the eastern Sierras. As a result, the U.S. Forest Service changed its plans for a drainage basin and Los Angeles took less water.

At the invitation of the World Bank, Botkin and Lee Talbot demonstrated that sustainable forestry claims had yet to be proven and made the case for testing proposed methods by computer simulation. The best practices would be carefully tried and monitored. Their paper gave the bank the evidence to direct its lending divisions to stop making loans to projects that encourage deforestation.

Private industry can also use ecology's insights to insure long-term survival. In Costa Rica, Botkin is working with a door manufacturer called Portico to put the company's forest lands on a sustainable-yield basis. The company has brought in forest scientists to study the natural fluctuations in forest growth and the relations among species in the ecosystem. Portico hopes the analysis and monitoring will establish a harvest pattern and evaluation process through which it can maintain a healthy forest as well as a constant supply of mahogany.

Since governments are practically the only institutions that can extensively monitor the global environment and deploy the necessary satellites, Botkin feels it is urgent that they catch up with ecological science. He points out, for instance, that satellite monitoring of oceans can provide information on the changes in currents and temperatures that cause big changes in fish populations. If fishing fleets had to use this information rather than abide by fixed quotas, we could approach sustainable management of our disappearing marine resources.

Botkin's work led him to propose a National Ecological Survey similar to the present Geological Survey. The agency that Botkin proposed would assume certain functions now scattered throughout EPA, NASA, NOAA (National Oceanographic and Atmospheric Administration), and other agencies to provide permanent monitoring of the environment and analyze its dynamics. Funding would come from user fees, charging for recreation, and the sale of natural resources. During his first one hundred days in office, President Clinton signed an order reorganizing scattered government research efforts into a single National Biological Survey. Clinton's proposal, however, is more limited than Botkin's. It focuses mostly on wildlife, and funding is dependent on political persuasion. An agency budgeted by the president and funded by Congress can easily find itself bound by a definition of nature that appeals to politicians, lobbyists, or the public. In fact, the National Biological Survey, with its mission to classify ecosystems across the country, was instantly perceived by property rights activists as a way to control private property through a national land-use plan.

Using new technology to redefine nature has always been a scientist's riskiest work. Industry, government, environmentalists, and even churches bet a lot of prestige and money on their vision of nature. In some ways, the philosophers of the environmental movement have functioned like the medieval church, opposing the idea that science can discover nature's secrets or manage the global environment. Many environmentalists argue for a hands-off policy on the grounds that we can never know how nature works. Thus we can only mess it up.

The fear of science often comes from sophisticated and trusted sources. A staff writer for the *New Yorker,* Alex Shoumatoff, wrote *The World Is Burning* to praise Amazon rubber tappers for protecting the environment and to raise the alarm about the species being lost there. Shoumatoff, in his rush to make his case in any way possible, ignored evidence of the rubber tappers' own destructive practices. (Three years later, rubber tappers turned against environmental groups whose support had been won by the late Chico Mendes, a rubber tapper and member of the Brazilian Communist party.) Shoumatoff reveals his fear of facts when he writes "numbers are a temperate zone precision trip" and when he calls hard data "the idlest of superstitions."[23]

Donella Meadows, coauthor of the Club of Rome's 1972 bombshell, *The Limits to Growth,* may have subscribed to the study's wildly wrong predictions in part because she cares so little for the perspective that historical fact lends to human behavior. For her, history seems to be whatever anyone says it was. In 1991, she told fellow nature writers, "I don't know what history says, I wasn't there."[24]

Many environmentalists also believe that power and corruption have pushed government scientists and technicians to environmental sin. Consequently, the new ecologists must follow in the tradition of Galileo, the first modern scientist to argue that we can know how the world works. Galileo, of course, was condemned by the church for his arrogance, had to smuggle his last works out of Italy, and died under house arrest. Modern ecologists are not risking their lives or even their jobs, but the conclusions we must draw from their work are not always welcome.

One of the most important conclusions to be drawn is that except for human beings, all of nature's creatures have been living on a very unpredictable planet where changes are often swift and devastating. If you happen to be a goose looking for a nest in an unusually cold arctic spring, or an anchovy in a suddenly cold ocean current, nature is neither compassionate nor wise. As far as we know, no creature but the human animal has been able to imagine things any differently. Now that we have fallen in love with the vision of a kind, balanced nature, we must learn how to intervene in nature's chaos to bring forth that vision. The changes we want in nature are quite grand and they will get grander as our population grows.

Managing nature to obtain the desired effect will have to be a hands-on effort. The great political question is, whose hands? The answer will depend on how well the environmental movement can catch up to and absorb ecology, and then how well it can communicate the message to the general public. When environmentalists talk about this final decade of the century as being a "make or break" time for saving the planet, they often talk of "converting" government, industry, and the public to a new way of thinking. Few of them realize that maintaining a healthy and beautiful planet may first depend on their own conversion:

> We talk about the spaceship Earth, but who is monitoring the dials and turning the knobs? No one; there are no dials to watch, only occasional alarms made by people peering out the window, who call to us that they see species disappearing, an ozone hole in the upper atmosphere, the climate change, the coasts of all the world polluted. But because we have never created the system of monitoring our environment or devised the understanding of nature's strange ecological systems, we are still like the passengers in the cabin who think they smell smoke or, misunderstanding how a plane flies, mistake light turbulence for trouble. We need to instrument the cockpit of the biosphere and to let up the window shade so that we begin to observe nature as it is, not as we imagine it to be.[25]

CHAPTER 8

Who Owns Nature?

The general rule at least is, that while property may be regulated to a certain extent, if regulation goes too far, it will be recognized as a taking.
—Supreme Court Justice Oliver Wendell Holmes

The public at large, rather than a single owner, must bear the burden of an exercise of state power in the public interest.
—U.S. Claims Court decision in *Florida Rock Industries v. United States,* quoting *Agins v. Tiburon*

My friend Stanley Selengut, who runs Maho Bay Camp in the U.S. Virgin Islands as an ecotourism resort, says protecting the environment pays. He does almost everything that environmentalists want builders, developers, and businesspeople to do—use recycled materials and natural energy sources, serve health foods, landscape with natural plants, use low-flow toilets and showers, avoid toxic chemicals, encourage the simple life, and more. His new up-scale experiment, Harmony, is built from reconstituted plastic, glass, newsprint, old tires, and scrap lumber. Electricity comes from solar panels and a windmill. And Stanley doesn't mind telling everyone that Maho Bay makes a 20 percent profit.

Stanley demonstrates the basic idea of environmentalist economics—doing good things for the environment pays off. I am happy for Stanley, but the broad generalizations that people make from his success do not hold up. Stanley succeeds in great measure because his resort appeals to environmentalists who believe what he is doing is right. In a similar manner, a church survives and grows because its believers fill the offering plate, not because its faith is better than another faith.

If doing what environmentalists think is good for nature always paid off, only stupid or masochistic businesspeople would do anything else. Environmental groups would not have opposed the 1993 North American Free Trade Agreement (NAFTA), because Mexico would have seen the dollar signs and adopted all the regulations that environmentalists insist on. All the farmers in Kansas would be growing organic wheat. The men's shops in Washington, D.C., would be selling natural tan cotton dress shirts instead of bleached white dress shirts. Owners of woodlands would be bidding madly for the privilege of hosting the spotted owl, the red-cockaded wood-pecker, or a grizzly bear. Homeowners intent on improving their invest-ment would be installing composting toilets.

Since the environmental movement began, it has supported thousands of regulations and ideas for environmental improvement. It is now clear that few of these lead to profit, and most take a big bite out of someone's earnings or the national economy. Yet it is not true to say that maintaining a healthy environment is incompatible with good economics.

What my friend Stanley seldom emphasizes and what environmentalists overlook is that if he were not making a profit, he would not do things that are politically correct for environmentalists. He might not run his resorts at all. The principle that Stanley Selengut demonstrates is one that environ-mentalists do not want to recognize: the profit motive works. But "profit" is too narrow. Stanley gets more satisfaction than just money. A better term for the principle that moves him is *self-interest*. Self-interest sounds like greed to affluent environmentalists who seldom have to think hard about where their meals or money come from. Their rigid faith in natural har-mony prevents them from accepting a universal principle of the living world. Two writers for the *Atlantic Monthly* recently suggested that the environmental movement "has set itself an unnecessary obstacle by largely ignoring the fact that human beings are motivated by self-interest rather than collective interests."[1] Unfortunately, the movement is programmed to break out in hives if the real details of self-interest are discussed for more than a few minutes. As I have shown in the history of the movement, its religious, philosophical, and economic foundations crumble if it admits the positive role of self-interest and its economic embodiment—capitalism.

If the 1990s really are critical years for the environment, it is because an alternative movement has begun to offer economically sound solutions to environmental problems. Since history has generally led the environ-mental movement in the direction of Socialist solutions, it is not surprising that the first organized alternative has embraced the powers of private property.

Environmentalists, who have a vested interest in seeming to be the only people who love nature enough to figure out how to save it, call the property rights movement reactionary. This is in part true. But the response to the politics of the environmental movement also draws its strength from historical forces that indeed have solved many of the world's most vexing problems, and it may be tapping parts of the human character that are more constructive than the fears and fantasies bred in the Romantic movement. Self-interest is something that environmentalists know firsthand.

Suppose we spin a Wheel of Fortune whose segments are labeled teachers, writers, coal miners, biologists, bus drivers, and so on. Whomever the arrow points to when the wheel stops will have to pay, say, $20 million or $30 million for the longleaf pine stands needed by the endangered red-cockaded woodpecker. The group picked by the next spin will pay for a few hundred thousand acres of grizzly habitat. Then lands for desert horned lizards, Puerto Rican crested toads, and Connecticut rattlers. Then wetlands, scenic river basins, mountains, greenbelts, and farmlands. Unfair, of course. But the system by which we choose who really does pay is just as crazy.

Back in the days when we wanted nothing more than a few well-defined parks and seashores, and when the national debt was considerably lower, government bought the lands with our tax money. Today the polls say we want a lot more, but not if we have to buy it. So we have taken what we want by law and regulation. The property owners from whom it is taken feel singled out by a Wheel of Misfortune.

Is this a case of the public good versus individual rights? Even assuming everyone agrees that the Connecticut timber rattler is a public good, and a greater one than moderate-income housing, is it fair to single out one citizen or a small group to bear the cost? This is the question behind the growing challenge to the environmental movement's methods. Vice President Gore says that we must make sweeping changes, but the more environmental change we want, the bigger the "taking" problem will become.

We have grown so accustomed to thinking that environmental measures are opposed by the callous and greedy and supported by the public in general that we fail to understand that this is not a problem of us versus them. If the American public has a subterranean bedrock unity, it is the right to enjoy private property—at least our own. I am reminded of Professor X, a specialist in Latin American studies, who believed strongly in laws and taxes to spread American wealth more evenly and to redistribute land. He believed in what people often call "economic justice." In the mid-1970s, when I was both president of a state conservation group and a realtor helping landowners plan environmentally sound development, Professor X came to me for help in dividing some lots from a one-hundred-acre farm he had purchased near town. When he saw my plan and the prices I suggested, he said, "Well, I thought I could get almost twice as much for this kind of lot."

"You can," I said. "But these prices give you more than a 50 percent profit, and I thought you were interested in seeing that they were available to the greatest number of people possible."

In those years, working near a university community polarized by the Vietnam War and other social issues, I discovered that when it comes to money, the country is not divided into liberals and conservatives, or environmentalists and exploiters. I went on to become a specialist in property values and an expert witness for scores of landowners whose private property was being taken whole or in part for public lakes, conservation easements, greenways, parks, footpaths, bikeways, and universities. And for environmental groups and governments who were doing the taking. I learned that only the rich and people near the end of their lives willingly give land away for nothing (although they usually take the tax deduction). In America, I discovered, Professor X is Everyman.

We own land for many different reasons, but we all agree that land is a great investment. And we believe that neither government officials nor anyone else ought to tell us what to do with our land. The exception is that they may tell us to do what we are already doing or what we want to do. In my case, let them limit my woodlands to forest uses. That's what I bought them for. All human beings want to possess the goods that give them power, status, or pleasure, whether those goods are stone arrowheads, a shaman's bag of herbs, a car, an oceanfront lot, or a house in Levittown. This important human truth has become a kind of glass ceiling for the environmental movement, or perhaps it is more like a wall of glass bricks.

Professor X doesn't chew tobacco and he can't tell a loblolly pine from a pond cypress, but he shares his basic American values with people south and east of here in the North Carolina coastal plain, where a lot of large and small landowners say they are suffering from "RCW." That's red-cockaded woodpecker, an endangered species. One man, whose father began managing more than seven thousand acres of longleaf pine for wildlife and timber, recently found out that his trees had grown so old and large that they had become RCW habitat. Dan Gelbert, one of North Carolina's most successful consulting foresters, says that the family's careful management has endowed it with some $2 million in timber that cannot be cut. Gelbert relates how a woman recently called and asked that he sell all of her growing longleaf pines before they get old enough to attract woodpeckers. And yet another client, an elderly farm couple, learned that they cannot cut their last stand of mature pines to pay their medical bills because they have RCW.

RCW is only one of many such problems across America. Government regulations—from local zoning to the Clean Water Act—have demanded changes in lifestyles and land use that usually impose real costs on landowners. The most appealing to the media and often the most galling to landowners are the rules written to protect endangered wildlife such as the red-cockaded woodpecker. Passed in 1973, the Endangered Species Act prohibits anyone in or out of government from "taking" a threatened or endangered plant or animal. This includes butterflies and minnows as well as grizzly bears and spotted owls. At the beginning of 1993, some four hundred species were protected. Whether they know it or not, taxpayers have paid along with private landowners. Federal and state governments spent $177 million in 1991, up from $43 million in 1989. Approximately one hundred species are added to the threatened or endangered list every year. As species are added, the cost to both taxpayers and landowners increases. Stories like those below multiply.

In California, Riverside County ordered all developers to stop work in an area that the U.S. Fish and Wildlife Service said was an important habitat for the kangaroo rat. The county devised a temporary plan to save the rat. It put 75,000 acres off limits to developers. Local developers who went ahead with work outside this study area had to pay a $2,000-per-acre fee to help protect the rat's future. These costs, of course, are eventually paid by home buyers.

When Emery Purslow, a retired merchant mariner, bought 146 acres of Maine coast in 1986, he planned to sell five-acre lots and use the profits for his retirement. He did not know that a pair of bald eagles had already chosen one of his lots. By the time the state Department of Environmental Protection and the Audubon Society finished estimating how much land the eagles needed to enjoy undisturbed peace, their lot had grown to about thirty-eight acres.

The hottest problem right now is the regulation of what the government defines as wetlands. Wetlands have become controversial for several reasons:

Wetlands

- Wetlands are the nation's most productive natural habitat.
- Coastal wetlands are usually in areas of high property values.
- At least half of the nation's original 215 million acres of wetlands have been greatly changed or lost.
- Seventy-four percent of the remaining wetlands are privately owned.
- The government's definition of wetlands includes lands that few people ever considered "wet." For the government, wetlands are places where the water table is within eighteen inches of the surface for one week of the year, or where any of some seven thousand species of indicator plants grow.

The major government wetlands program is a monument to bureaucracy, and an invitation to rebellion in a country with a strong tradition of private property. Most of the controversy started in 1972, when Congress amended the federal Water Pollution Control Act (renamed the Clean Water Act in 1977) to include all waters of the United States. After years of government toleration for industries and cities using rivers as sewers, a Ralph Nader study group staffed mainly by graduate students turned out a catalog of abuses entitled *Water Wasteland.*[2] The book whetted the environmental movement's appetite for revenge. In 1975, the Natural Resources Defense Council won a court order applying the law not just to waterways, but also to swamps, prairie potholes, marshes, and other lands occasionally soaked. The court gave the green light to extend the definition of wetlands to "all waters of the U.S. . . . to the maximum extent permissible under the Commerce Clause."[3] Since then, both the U.S. Corps of Engineers and EPA have argued in court that a wetland can be anywhere a waterbird crossing state or national boundaries stops to rest. In 1985, Senator George Mitchell, a Democrat from Maine, added the "glancing goose" clause, insisting that waters covered by the interstate commerce clause of the Constitution, and therefore by the Corps of Engineers' regulations, are any waters that are used or *could be* used by migratory birds. In August 1993, President Clinton proposed changes to exempt existing farmlands and to add some flexibility to the permit procedures that anyone proposing to work in a wetland area would have to follow. The changes pleased no one, and the wetlands regulations continue to give government officials power over a vast area of America's private lands.

The arguments over how to define a wetland became so hot that in 1989 several federal agencies developed the *Federal Manual for Identify-*

ing and Delineating Jurisdictional Wetlands, commonly known as the "Wetland Delineation Manual," to guide regulators. According to the manual, a piece of ground dry to the touch except in rainy weather can be classified as a wetland if it is wet within eighteen inches of the surface for at least one week during the year and if it grows plants associated with wet soils. (They could be plants that grow equally well in uplands as in wetlands.) The list of wetlands-defining plants includes some seven thousand species. Attempts to revise the manual or produce a new manual failed as the Bush administration came to an end.

How far the new definition expanded wetlands regulation is illustrated by looking at old and new estimates of wetland coverage. In 1956 conservationists estimated that 75 million acres remained. They calculated this as 58 percent of the original 130 million acres. Then in the late 1980s, while lamenting the continuing loss of wetlands, environmentalists estimated that some 100 million acres remained from an original 200 million. In little more than thirty years, wetlands somehow expanded by almost 25 million acres while destruction continued. Then the 1989 manual redefined wetlands and another 100 million acres appeared, most of it privately owned. The United States suddenly had 200 million acres of wetlands again. In short, through miscalculation and redefinition, wetlands have more than doubled since 1956.

Environmentalists condemned President Bush in the 1992 campaign for abandoning his pledge of "no net loss of wetlands." Yet his agencies, using standards supported by environmental groups, actually increased the amount of officially defined and regulated wetlands. Clearly the environmentalists' goal is not to define wetlands with scientific accuracy. Their goal is to stop development and limit the rights of landowners to activities they approve. Ironically, despite all the regulations, many ecologically important wetlands *are* disappearing (some from natural causes such as storms and erosion), especially in popular coastal areas, in the southern coastal plains where timber and farm crops thrive, and in the upper Midwest's great migratory pitstop, the Prairies Pothole area.

In the real world, the important wetlands are a modest fraction of the lands covered by the expanded regulations. In addition, many of the most important lands could never have been developed, so the regulations were not really necessary. Others could have been preserved by outright purchase or by buying a conservation easement, that is, a guarantee that the owner will not disturb the wetlands. In their belief that greed and a free market are equivalent, environmentalists have missed the single great-

single greatest cause

est cause of wetlands destruction. The cause was not the free market. Rather, it was government subsidies—the reason for so much environmental destruction in many Socialist countries as well. During the hungry years of the Depression, the U.S. government offered landowners money to drain lands that could grow crops.

In 1965, before the Clean Water Act passed Congress, Supreme Court Justice William Douglas proposed a Wilderness Bill of Rights to "ban the draining of any wetlands until and unless the appropriate conservation agency of the state or of the federal government is given the opportunity to acquire it and hold it for a conservation objective."[4] The first part of his proposal—to ban the draining of wetlands—has been realized piece by piece over the years, while the suggestion that government acquire instead of regulate has disappeared. The result is that present wetlands regulations position the government to force individual landowners to pay for what it believes is a general public good. In the words of the regulators, "When the protection of natural, cultural or aesthetic resources or the assurance of orderly development are involved, a mere loss in land value is no justification for invalidating the regulation of land use."[5]

By assessing the costs of preservation against scattered individual landowners and by extending government authority unnecessarily, environmentalists have ignited a war of principles that may rival that sparked by taxation without representation.

Supporters of regulation like to point out that under the Bush administration, the present wetlands protection program denied only 3 to 4 percent of some 15,000 permit requests. That was still six hundred angry landowners. And many more were discouraged by the cost of permits and the threat of bureaucratic red tape from requesting a permit. A detailed study of permit applications by a former chief of regulatory with the Corps of Engineers found that in 1990 and 1991 more than 40 percent of applicants withdrew in frustration. The relatively small number of legal actions brought by the government does not make the law any less objectionable to landowners. (Would anyone defend a law requiring separate drinking fountains for blacks just because it was applied only a few times a year?) However many times the law strikes a landowner, it has established the Corps of Engineers and EPA as national zoning authorities. To suggest that lack of enforcement makes the law acceptable is like saying that a man waving a loaded gun in a crowded room should be ignored. Other groups have vehemently protested much more ancient and unenforced laws—those against sodomy and cohabitation, for example. The damage from regulatory laws is spreading, not diminishing. The growing number of cases in which well-intentioned property owners have been wounded does not inspire confidence in the government's choice of targets. In addition, the Clean Water Act was passed to control pollutants that could move from one waterway to another, not to protect isolated or temporary standing water, or even to protect wetlands. In fact, the act never mentions wetlands at all, citing only discharges into "navigable waters of the United States." The property rights advocates say, if Congress wants to protect wetlands, it should say so clearly in a law.

The wetlands regulations have resulted in often absurd rulings. The federal government indicted a Colorado farmer who redirected an artificially channeled river back to its natural bed, and it denied a Wyoming woman a permit to plant roses on her land. Other landowners have gone to jail for building duck ponds. According to the government, so many ducks were attracted to the ponds that their droppings polluted the water. EPA used Senator Mitchell's "glancing goose" clause to slap a $50,000 fine on a Chicago-area home builder. Hoffman Homes proposed building houses in an old cornfield in 1985. The local government required Hoffman Homes to fill in a shallow depression of less than an acre as part of a park construction program. EPA sued, charging that a passing bird might have used the area if a rain left temporary standing water.

The situation has not progressed to the point where a landowner who hangs a birdfeeder or builds a duck pond is going to have to provide flush toilets for wildlife, but the zany zealousness of some enforcers, combined with the potential for economic ruin, makes for frightened and angry landowners. Landowners who fear the results of protecting wildlife and who are angry at the people who tell them how to do it make lousy conservationists. Like every member of every species, their first priority is to protect their habitat.

All landowners have choices to make about how to manage their land. The present situation encourages them to manage it so that it does not attract or support endangered wildlife. The woman who called the forester to cut down her pines before the woodpeckers came made a rational decision for her own good. Others simply ignore the law. They tell the foresters, "If you see any woodpecker holes, don't tell me." In the Pacific Northwest, many landowners summarize their strategy as "shoot, shovel, and shut up."

When Secretary of the Interior Bruce Babbitt proposed the National Biological Survey (NBS), he recognized that previous regulations were his greatest handicap. His solution was to ask that discussions of NBS be secret and exempt from the Freedom of Information Act. When Congressman Charles Taylor, a Republican from North Carolina, asked him why he wanted secrecy, Babbitt replied that information on which species are considered endangered might frighten landowners into getting rid of the liability or the habitat that supports it.

Perhaps Babbitt's fears were justified, even if his tactics weren't. The August 1993 issue of a widely distributed property rights newsletter called the NBS proposal "the Pending 'Mother' of All Land-Use." The past abuses of government power over private lands have made NBS a bogeyman. Property rights activists have decided to oppose the gathering of important information because they are sure that whatever scientific value it may have will be secondary to its use in expanding the costly regulation of private property:

> It is ironic that the Constitution explicitly forbids the government from requiring that a citizen quarter a soldier (provide food and shelter for a soldier) but the government can now require the same citizen to quarter a grizzly bear, a spotted owl, or any other member of a species declared to be threatened or endangered.[6]

REGULATION AS TAKING

Environmentalists overwhelmingly back regulations and discount the costs. When people object, environmentalists too easily assume that they are driven by greed and callousness, even though their objections stem from the same pride and economic expectations I have seen among environmentalist landowners. Instead of searching for solutions from common experience, the fight has become almost as polarized as the abortion debate.
The biggest environmental groups, including the Audubon Society and National Wildlife Federation, admit no virtue or honor among property rights activists. The federation has opposed laws that would require a state to review regulations to determine if they might trigger expensive lawsuits about taking property. The federation is convinced that people who favor these laws have a hidden anti-environment agenda "under the guise of preserving 'property rights.'" An article in the federation's newspaper for its affiliates calls people fighting to receive compensation for lost values "anti-environmentalists." The writer worries that new state laws to protect property owners will "result in major delays of needed regulations by requiring an expensive analysis and a lot of red tape."[7]

This may be true, but it hardly wins any support from property owners to suggest that delays and expensive analysis are acceptable in something such as an environmental impact statement but not when protecting private property rights and perhaps a family's life savings. This attitude leads many property owners to think of environmentalists as antiproperty. Environmentalists like to say that the choice is between self-interest and public interest, but if I want my roadside to look natural to please my eye, or a wilderness where I can see grizzlies, isn't that also self-interest?

Just this kind of blindness turned the Audubon Society lobbyist Ann Corcoran of Maryland into one of the property rights advocates whom the National Wildlife Federation calls an anti-environmentalist. Corcoran has a forestry degree from Yale University and worked for Audubon from 1976 to the late 1980s. Her husband works as an attorney at EPA. She became a property rights activist when the National Park Service started what she says were secret maneuvers to incorporate her home and land in the Antietam National Battlefield. The Corcorans, like many of their neighbors, have gone to great expense to restore their home, a former Civil War hospital. She is infuriated that a government agency would try to take her property.

In one of the most celebrated wetlands cases, a man who has a solid record in conservation was sent to jail. Bill Ellen and his wife run a rescue and rehabilitation program for injured waterfowl and wildlife. In 1987, Ellen, a marine engineer, was working for the millionaire futures trader Paul Jones II, turning his 3,200 acres of Maryland farmland into a wildlife preserve. With the assistance of the Maryland Department of Natural Resources, Ellen began work on a showcase 103-acre sanctuary of ponds, wildlife food plots, and grasslands for ducks and geese. Somewhere in the process, Jones pleaded guilty to the misdemeanor of negligent filling of wetlands and paid $2 million. Ellen decided to fight the charges in court.

Ellen argued that he and Jones had indeed filled a forested area that the government considered wetlands, but the area had been so dry that he had to wet it to suppress the dust during clearing. Also, the land did not fall under the wetlands regulations when he began work in 1988, but was included only when the official definitions changed in 1989. Besides, he had replaced the area with new ponds.

An EPA scientist testifying for the government said that the ponds degraded the environment by attracting too many birds. Too many birds make too many bird droppings. Before Ellen's work, the droppings stayed on the "wetlands," the scientists testified, but afterward they were spread about in the water.

The "expert" answer puzzled Judge Frederick Smalkin, who asked, "Are you saying that there is pollution from ducks, from having waterfowl on a pond, that pollutes the water?" The jury convicted Ellen and sent him to jail for six months. In June 1992, he returned home, a hero of the growing property rights movement.

Peggy Reigle says that people such as Corcoran and Ellen are the vanguard of a third wave of environmentalism, which will recognize that Americans' high regard for private property is not the cause of environmental problems but the core of the solution. Reigle retired as vice president of finance for the *New York Daily News* to enjoy the beauties of Maryland's Eastern Shore. When wetlands regulations threatened to wipe out the savings her neighbors had invested in land, Reigle became a property rights activist. In July 1990, she promised her husband her fight would be over by Thanksgiving. By 1993, she was chairing the Fairness to Land Owners Committee (FLOC) in Cambridge, Maryland, with some 13,000 members in forty-six states. She said she was working twelve hours a day, seven days a week, without pay.

Because business pays the most for environmental regulation, property rights activists often find their cause joined by oil companies, labor unions, and the Farm Bureau. Their opponents would like nothing more than to tie them to big business. Environmentalists sometimes try to equate property rights groups with the wise use movement, which is heavily loaded with western ranchers, large farmers, mining and timber companies, and other extractive users. The amorphous wise use movement embraces the goals of the property rights movement. However, at the Environmental Grantmakers Association conference in October 1992, representatives of more than 130 corporate foundations, including Apple, Chevron, and Ford, heard a surprising report on this relationship. The report was sponsored by the W. Alton Jones Foundation. Its author, Debra Callahan, after surveying all fifty states, concluded that the property rights movement is not allied with the wise use activists in the West and its members are not the foot soldiers of corporations. The wise use movement's focus is on rights to public lands and resources—rangelands, mineral deposits, rivers, and forests. The property rights movement, Callahan reported, focuses on government action that affects private lands.[8] Callahan told the meeting, "This is pretty much generally a grass-roots movement." These roots, the diversity of the groups involved, and their modest means, Callahan con-

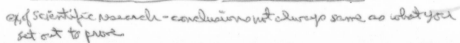

cluded, make discrediting them difficult. Compared to the big environmental groups, the property rights groups are poverty-stricken. Searching for a way to help environmentalists discredit the property rights movement, Callahan suggested trying to link it to extremist groups such as the John Birch Society and Lyndon LaRouche. As the editor of the environmental magazine *Buzzworm* said:

> Now that government may be taking the environmental crisis of our planet seriously, the eco-bashing rhetoric of the radical Right takes on a new light for the common man or woman whose job is indeed threatened, for the developer whose plans will be stymied, or for the industrialist whose company will be penalized—all by environmental restrictions.[9]

There is no evidence that property rights activists love nature less than other Americans. And they may be fighting for a tradition that is even more basic to the American psyche and system of government. Environmentalists who plan to pit environmental regulations against property rights may lose more than they gain. Callahan told the Environmental Grantmakers Association that the property rights movement would seize the high ground if the environmental movement continued to be perceived as composed of affluent, white-collar elitists. "We are the enemy as long as we behave in that fashion," she warned.[10]

With Gore and Clinton supporting stronger and more costly environmental regulations, the taking issue will become an ever larger obstacle for the environmental movement. Across the land, more and more environmental initiatives are being taken to court over property rights. The trend began to gather momentum in the mid-1980s. When California's floodplain zoning left a Lutheran church with nothing but picnic property, the U.S. Supreme Court ruled that when government leaves a property with no commercial uses, it must pay. This seemed to be little more than a slight expansion of the 100 percent damage rule, but it was quickly followed by a case that echoed in the planning community across America.

When the owners of a coastal property tore down an old house and asked permission to build a larger one on the site, the California Coastal Commission approved the proposal, but the owner was required to dedicate a third of the property to public use. In 1987, the U.S. Supreme Court ruled that the requirement had nothing to do with the commission's legal authority to protect the coast.

This decision had to be in the mind of Maryland's secretary of natural resources when he found out that the state's Department of Environment had been demanding deed restrictions unrelated to wetlands protection. In one case, a landowner who wanted to build his retirement home on a half acre of his eighteen acres was told that in return for a permit, he would have to record a deed restriction prohibiting the development of any more land. Soon after Peggy Reigle confronted the secretary with this situation, the governor ordered an end to the practice.

In the mountains of Virginia, owners of riverside land have claimed the right to control fishing and hunting under the terms of land grants made in the eighteenth century by the English kings George II and George III. Virginia has traditionally recognized these titles, but the landowners along one popular river are now in court claiming that a deal between the U.S. Corps of Engineers and the state to make the river a public fishery takes their land without payment.

In Long Beach, New Jersey, the U.S. Court of Federal Claims awarded the developer of Loveladies Harbor $2.66 million when the Corps of Engineers denied a permit to build houses on 12.5 acres it claimed were wetlands. On the same day in 1990, the same judge decided that a limestone mining company in Florida should be awarded more than $1 million because Corps regulations had stopped mining on 98 of its 1,560 acres in order to protect ground water reserves. The judge said that the evidence had not proven that the company's operations caused water pollution.

In 1991, the courts levied the largest damage award in history against the U.S. Department of the Interior. The U.S. Claims Court and the U.S. Court of Appeals told the department that by barring a coal company's operations in a protected area, it had taken property rights worth more than $150 million. The U.S. Supreme Court heard the government's final appeal and confirmed the verdict.

In July 1993, the U.S. Court of Appeals reversed the $50,000 fine on Hoffman Homes for filling a low spot where migratory birds might stop. The court wrote:

> It is true, of course, that migratory birds can alight most anywhere. As Gerald Bade [of the U.S. Fish and Wildlife Service] testified, he has seen mallards in parking lot puddles. . . . The migratory birds are better judges of what is suitable for their welfare than are we. . . . Having avoided Area A the migratory birds have thus spoken and submitted their own evidence. We see no need to argue with them.[11]

Cases that are not directly related to environmental regulations are also setting precedents that will have a profound impact on the countryside. In

1985, a federal court handed down a rare award for property damaged by changes in zoning regulations: when Dubuque, Iowa, rezoned an industrial park property as residential, the Eighth Circuit Court awarded its owner $89,812. If downzoning is considered a taking, isn't it true of environmental regulations as well?

By 1992, nearly two hundred taking cases were pending in the U.S. claims courts.

For those of us who are deeply interested in preserving open spaces, this idea of compensation could be ruinous.[12]

We can't protect our nation's environment if we are going to give in to private property rights.[13]

States have tried to argue that well-established law allows them to take property without payment in order to protect the public from a nuisance or an obnoxious use. The first big blow to this defense came in 1988 when South Carolina told a well-heeled young developer named David Lucas that he could not build on two oceanfront lots for which he had paid $975,000. The state said the prohibition was necessary because a new law specified that houses must be a certain distance back from the water, and Lucas could not meet this building requirement.

Lucas sued, enlisting help from a variety of experts, including the University of Chicago's Richard Epstein, whose 1985 book, Takings, turned up the heat on the debate over regulatory taking. When the case reached the U.S. Supreme Court, the justices voted 6–3 that the state could hide behind the nuisance defense only if it could show that it was grounded in time-honored common law. Last year, the South Carolina Supreme Court decided that the state could not declare Lucas's proposed home a nuisance unless it required all his neighbors to tear down their homes. Since the state had raised no objection to Lucas's neighbors building on similar land, the court ordered the state to pay Lucas or let him build. The state paid Lucas.

Then the state decided to resell the property. The irony was not lost even on John Echeverria, chief counsel for the Audubon Society. Echeverria said that the state's decision appears hypocritical "when it is willing to have an economic burden fall on an individual but not when the funds have to come out of an agency's budget."[14]

problems

The Natural Resources Defense Council lawyer Albert Meyerhoff probably exaggerated when he said that Lucas's victory would mean "open season on environmental laws,"[15] but it does signal that government cannot blithely change the rules after property owners have bought land. A little-noticed footnote in the Lucas opinion may also be important: in footnote 7, the court says that it makes no difference what other properties an owner might hold, even nearby.[16]

Dozens of important cases are working their way toward the higher courts. If the trends cited here continue, making regulations is going to become more and more expensive. Court cases like these cost millions of dollars. Just as the civil rights movement put legal teeth in the Bill of Rights, property rights activists have proposed dozens of state and federal laws to put teeth in the Fifth Amendment's "just compensation" clause. By 1993, more than twenty state legislatures had considered assessment bills or compensations bills. Assessment bills define when a regulation that damages property values requires payment. Compensation bills require payment if the damage reaches a certain threshold. A bill sponsored by a third of the Florida House would have required compensation if a regulation reduced a property's value by 40 percent or more. The National Wildlife Federation, much to its surprise, helped kill a Wyoming compensation bill, but Delaware passed the first such bill in 1992, followed by Arizona and Utah. For the past three years, the Idaho legislature has sent a bill to Governor Cecil Andrus. The governor, a former interior secretary under President Carter, has vetoed it three times. In Congress, the Senate minority leader Bob Dole gathered twenty-one Republican sponsors for a taking bill. Dole planned to introduce his bill as an amendment to the bill making EPA a cabinet-level agency, but withdrew it under fire from Vice President Gore. In the House, the California Democrat Gary Condit's bill had sixty-one cosponsors. House Republicans, under the leadership of California freshman Richard Pombo, a thirty-two-year-old farmer, formed the Congressional Property Rights Task Force to hold hearings and draft new laws.

Environmentalists often oppose new property rights laws much the way segregationists once opposed civil rights laws, saying that state and federal constitutions already provide enough protection. Peggy Reigle said of Maryland's proposed legislation, "This is the Civil Rights Act of 1993."[17] Property rights activists, like civil rights activists, are winning their early battles with existing law. Most of these battles have such names as *Lucas v. South Carolina,* or *MacDonald, Sommer, and Frates v. County of Yolo and*

City of Davis. The results of these legal battles are still indecisive, but the property rights groups have opened a second front in state legislatures. The rise of the environmental movement did not give birth to a new legal debate. Rather, it has become the catalyst for resolving an ancient dilemma.

A BRIEF HISTORY OF PROPERTY BATTLES

The debate over public and private property rights has been going on for more than two thousand years. The Romans were the first to preserve a code of law recognizing private property rights. They recognized that property means nothing without protection against government encroachment. Under Roman law, the government could not take property without paying a sum determined by special judges. However, as early as the first century B.C., the Romans also recognized the right of government to regulate. An owner could not build so close to his neighbor or so high that he blocked a valuable view or light.

Churchmen in the Middle Ages also recognized the power and danger of private property. The influential St. Thomas Aquinas insisted that the public good provides the ultimate reason for holding private property. This continues to be the moral justification for government's power to regulate, and even take, private property. In twelfth-century London, where fire was a grave threat, the city council required adjoining neighbors to each give one and a half feet of land and to build on it a masonry wall three feet thick and sixteen feet high. In the countryside, the king enjoyed the privilege of declaring the trees and game in certain forests reserved for the throne, and therefore almost useless to the local community or baron, the nominal owner. Kevin Costner, who has played a number of anticapitalist roles including Robin Hood, might be dismayed to know that the legend of Robin Hood celebrates the struggle of citizens to win property rights from their government.

The chain of legal documents that eventually led to American property laws began in 1215 in England with the signing of the Magna Carta by King John. Until that time, property was not so much owned as "held." (Thus we still speak of *landholders.*) The king granted his nobles rights to hold manors so long as they pledged loyalty and would defend him. King John invited an attack on this royal privilege by taking property in order to pay for his disastrous wars. With the Magna Carta, or Great Charter, the barons extracted from him a stronger version of property rights suggested by shorter, vaguer charters that dated back before William the Conqueror. One of the chief grievances against the king was "arbitrary infringements of personal liberty and rights of property." The words in article 39 of the charter, which contain ideas similar to our own Constitution, guarantee that "No freeman shall be arrested, or detained in prison, or deprived of his freehold, or in any way molested; and we will not set forth against him, nor send against him, unless by the lawful judgment of his peers and by the law of the land." In article 28, the Magna Carta says, "No constable or other bailiff of ours shall take grain or other provisions of any one without immediately paying therefore in money." The Magna Carta also put an end to royal regulation of forests.

Reinforced by philosophers and by the role of private property in creating an individualistic middle class and unprecedented prosperity, these attitudes became part of revolutionary America. Similar ideas appear in the Constitution in the Bill of Rights, at the end of the Fifth Amendment: "No person shall . . . be deprived of life, liberty, or property, without due process of law; nor shall private property be taken for public use, without just compensation." For James Madison, who drafted the provision, it insured that people would not use the political process to do what my friend Professor X would like to do—use law and regulation to take wealth from one person and give it to another. It also insured the economic growth spurred by new settlers and by veterans who received land grants as pay.

Environmentalists who see property rights activists simply as greedy landowners and as representing big business have a shallow view of the sources of their opponents' passion, logic, and strength. The National Wildlife Federation attorney Glen Sugameli, for instance, has said that the property rights movement comes "directly from the radical agenda of Ed Meese [President Reagan's attorney general]. Commercial interests were cynically exploiting property owners' concerns."[18] The Maryland state senator Gerald Winegrad and the Audubon Society legal administrative assistant Susan Murray have made the same charge. Behind it are the memoirs of President Reagan's solicitor general, Charles Fried. His actual text says that Meese wanted "to use the takings clause of the Fifth Amendment as a severe brake upon federal and state regulation of business and property."[19] This may sound like conspiracy to some, but to people whose life savings and major investments are threatened, it sounds like common sense. Every good lawyer knows that if you do not understand your opponent's point of view, you increase the risk of losing.

However good the cause, over simplification misses the depth of the property rights movement and diverts attention from the key fact: a regulation or the acquisition of land almost always redistributes wealth. Environmentalists like to say the public good has taken precedence over private profit. But that evades the question of why the public hasn't paid for the damage.

The great fight over preserving wetlands and endangered species is not about whether we should do it. Americans agree on the value of wildlife. The disagreements are about what a wetland is and whether every square foot requires saving, and about which species or subspecies should be protected, whether they are really in danger, and whether the regulations that cost private property owners so much really work. And of course there is the question of who should pay. Private property activists (and most environmentalists when their own property is taken) would agree with the U.S. Claims Court in Florida: "The public at large, rather than a single owner, must bear the burden of an exercise of state power in the public interest."[20]

The history of population shifts is partly to blame for this war. As more and more Americans left the countryside for towns and cities, we became a nation divided between the rural few and the urban and suburban many. When environmentalists talk about Americans being detached from nature, they could easily be talking about themselves (especially since their headquarters are usually in Manhattan or Washington, D.C.). Why is it that environmental groups draw most of their support from city and suburb? If living close to nature gives human beings a special knowledge and a special appreciation, why don't environmentalists pay more attention to rural Americans who don't happen to be Indians? While farmers, ranchers, fishermen, and lumbermen may not always be good conservationists, neither are they detached from nature. There are bad ranchers and loggers, just as there are bad computer companies and car makers, but most have to pay close attention to the environment as well as to economics. If environmentalists are serious about harmonizing jobs and the environment, shouldn't they be paying more attention to the people they habitually call anti-environmentalists?

Regulations generally affect rural property or the fringes of suburbia more profoundly than urban areas. When Maryland considered its Private Land Rights Protection Act, Republicans and Democrats from city and suburb united in opposition. Their arguments in favor of regulations were also arguments for government to help them get something they wanted, but something owned by another citizen. The 125,000 people who own private land within New York's Adirondack Park discovered something similar soon after the state imposed a strict zoning plan to protect surrounding public land. Half of their land was zoned to allow one house on every 42.7 acres. Most landowners love the park but hate the regulations instigated by the Rockefeller family and legislators from cities and suburbs. The supervisor of Indian Lake, a rugged mountain climber and woodsman, said, "It's perfectly simple. The state couldn't afford to buy all the private land, so they zoned it."[21] The state now not only says how much can be built, but also where and what it must look like. One man who built a deck overlooking a river was charged by the state with building its railing too high. Mario Cuomo, the governor of New York, said, "Yes, we have taken away some of the rights of the people living in the Adirondacks, but that's the penalty they have to pay for living there."[22]

When the Sierra Club member Tony D'Elia arrived in the Adirondacks in 1969, he bought 3,500 acres to develop gradually. Building roads and putting in sewers, he invested a half million dollars a year. Then the zoning plan arrived and he had to respond to a seventy-two-page environmental impact outline. He spent six months and $50,000, but his environmental analysis did not satisfy the authorities. He spent another $75,000. When his plans were finally approved two years later, regulators attached sixty-two conditions, including a water-monitoring program that would cost $100,000 a year. D'Elia's mortgage holders foreclosed. Today development on private lands is possible only for big developers such as the ski resort operator who spent $1 million on plans to meet state requirements. Homes at the resort are more than twice the price of other homes in the area, and meals at the local lodge sport New York City prices. Prices and values have increased mainly to satisfy the demands of environmentalists who do not live, work, or invest in the Adirondacks. Adirondack Park property owners are convinced that the regulations they shoulder have transferred the value of their lands to people who could afford to pay but don't.

HOW TO DEFINE THE PUBLIC GOOD

Of course, regulations were not invented by environmentalists to steal property. Regulations take advantage of ancient concepts of ownership. We may say that our homes are our castles, but somewhat like feudal nobles, what we own is not absolutely ours. Ownership is not a legal rock, but something like a bundle of sticks. Few of us own every possible right to our property. We usually buy a house or land with a few rights, a few sticks missing from the bundle. The right to build a factory or operate a gas station has been removed by zoning. The right to divide the house into two apartments has been removed by restrictive covenants. We can sell a stick or two of our ownership rights. Some people sell oil or gas rights. My neighbor recently sold timber rights. I have given a neighbor the right to build a drive over one corner of my land—a "right of way."

The bundle of rights concept can make life very complicated when public and private sticks come into conflict. Take the case of the cattleman Wayne Hage. When armed U.S. Forest Service officials rounded up his cattle, charging that Hage had overgrazed his public range allotment, Hage sued the government for $28.4 million in damages for rendering his two-thousand-head operation unprofitable. Hage claims that his ownership rights predate the creation of national forests. The Forest Service says that grazing cattle on the public range is not a property right like a lease, but a privilege that can be withdrawn at any time, something like closing a road for repairs.

Take the contention of the former EPA administrator Bill Reilly, who criticized "mainstream attitudes about private property and freedom of action" for creating "tawdry gateways to some of our greatest national parks." Without regulations, he said, "the market will decide which land gets developed and which remains as it is."[23] He was quite right about tawdry development, as anyone knows who has been to the totem pole–littered Indian town of Cherokee on the east side of the Smokies. And at the western gateway of Gatlinburg, you can visit Dollywood, see Buford Pusser's death car, or witness the world's smallest evangelist. But is the government entitled to take away without payment the property rights of the Cherokees or the owners of Gatlinburg's amusements in order to maintain the value of the adjacent national parks?

The taking controversy is not over rights long gone or voluntarily relinquished, but over rights taken by government without just compensation. It is well and good and scientifically sound to say that preserving the red-cockaded woodpecker, the grizzly, wolves, clear water, or the acid balance of a lake protects the environment that supports us all, but it is also true that private property rights sustain us all, including the freedom we have to be environmental activists. (It is no accident that the environmental movement developed in capitalist democracies where private-property owners have recognized that basic freedoms nourish each other.) To take a valuable right without paying does not seem very different to the owner than if the government singled him out for a special individualized tax to save the red-cockaded woodpecker or a canoeist's view of the New River's banks. Environmentalists like to argue that what landowners lose is eventually compensated for by a healthier economy and other public benefits. Perhaps, but why are a few individuals chosen to make sacrifices to improve life for everyone else?

In 1992, the U.S. Supreme Court struck hard at the right of environmental groups to press their views on other people. Defenders of Wildlife had sued to force the U.S. Department of the Interior to consider endangered wildlife when it spent U.S. funds on foreign projects. In *Lujan* v. *Defenders of Wildlife*, the court said that environmental groups could not sue simply because their members might study the foreign animals or enjoy them on foreign tours, or even because the species are part of a nearby ecosystem.

The environmental columnist Alston Chase says that "as a political issue, property rights is a dinosaur." Farmers and other landowners are making the error of "believing their quarrel with environmentalism is about the economy and the Constitution. It is not. It is a dispute over values."[24]

"But the Constitution deals precisely with the means to resolve such value conflicts," responds Richard Stroup, a senior associate at the Political Economy Research Center, where scholars are developing a variety of free-market solutions to environmental controversies. Stroup divides government regulations into two types. The first "keeps landowners from violating others' rights by, say, polluting streams." While these regulations often impose large costs for small benefits, Stroup finds a second kind of regulation more irritating, that being "when certain landowning individuals are required to provide public goods without compensation."[25]

One reason the government cannot pay for its regulations is because the regulations try to do too much. The Endangered Species Act assumes that every species must be protected, and not only every species, but their habitats as well. The expanded interpretation of the Clean Water Act makes every standing puddle sacred. Edward O. Wilson, the famous Harvard zoologist and the author of *Sociobiology* and *Biodiversity*, argued in a 1991 article in *Science* that the first step to saving endangered species

> would be to cease "developing" any more relatively undisturbed land. ... Ending direct human incursions into remaining relatively undisturbed habitats would be only a start. The indispensable strategy for saving our fellow living creatures and ourselves in the long run is ... to reduce the scale of human activities.[26]

The reduction for which he begs is usually accomplished by government regulation, but it is also almost always a reduction in someone's livelihood or property value. The costs ricochet through society. During the Reagan and Bush years, environmentalists tried to soften their elitist image by sympathizing with the problems of the poor—the small farmer, children in the ghettos, migrant workers. The rhetoric continues and so do the burdens on the poor. When environmental regulations combine with inflexible building codes and social engineering, for example, the costs of home building go up. The people at the bottom suffer first. Many go without housing they might have had. Others rely on subsidies. Among builders and developers, no one builds low-cost housing without subsidies. The subdivisions that Levitt built after World War II to house hundreds of thousands of low-income veterans would be impossible today. The subsidies for low-cost housing are another way taxpayers (often low-income) foot the bill for environmental regulation.

I know, because I watched it happen in a community I was developing. I planned the development both to save the wildlife habitat and keep acreage tracts affordable. (Three acres, for instance, cost less than a quarter-acre lot in nearby communities.) My partners and I built a six-acre lake for recreation and wildlife—no motors, no individual docks, no clearing to the shoreline. Although we restricted the entire flood plain downstream to the nearby river, we nevertheless had to comply with engineering requirements designed for gravelly mountain soils (ours was piedmont clay) and developed areas. We had to drill a series of test wells along the back side of the dam, even though the state said it never sent inspectors to monitor the wells for seepage. The work cost tens of thousands of dollars in unnecessary engineering and construction, which we added to the price of the lots. For every thousand dollars added to a lot, several potential buyers were priced out. In miniature, this is what happens with all regulations that affect building and housing. Some of the costs are necessary. Many are not. Environmentalists who hold that all of nature is sacred seldom want to discuss the difference. In cases where costs are passed on, the developer or landowner loses nothing of value because of the government regulations, but the buyer loses money—and sometimes even the opportunity to own a home.

It would be nice to assume that we can save everything, but choices have to be made. The same year that Epstein's manifesto against regulatory takings inspired the property rights movement, the Duke University economist Robert Healy published *Competition for Land in the American*

South, which suggested that planners do more to preserve "unpriced values" such as wildlife habitats and ground water.[27] However, even if these resources have no price, protecting them does.

THE PROBLEM AS THE SOLUTION

"So you want to divide all the money there is and give every man his share?"
"That's it. Put it all in one big pile and split it even for everybody."
"And the land, the gold, silver, oil, copper, you want that divided up?"
"Sure—an even whack for all of us."
"Do you mean that to go for horses and cows?"
"Sure—why not?"
"And how about pigs?"
"Oh to hell with you—you know I got a couple of pigs."[28]

More than ten years ago, the Natural Resources Defense Council's sharpest thinker, Peter Borelli, writing in *Amicus,* warned that environmentalists had "ignored certain basic facts of life in America." As he laid them out, they were:

Fact: Land and personal freedom are strongly associated in the minds of most Americans.

Fact: Our system of land law since precolonial times has evolved and revolved around the preservation of social and economic rights associated with land ownership.

Fact: Our economic system is based upon a free-market approach to land.

Fact: Land is viewed as a commodity, both as a matter of self-interest and national tradition.[29]

Borelli said that extremists had not caused the repeated defeat of better land-use laws: "The real cause has been the inability of reformers to deal with rather than confront prevailing attitudes about the land."[30] For most environmental groups, the self-interest of property owners still seems like an unmovable obstacle. Jay Hair, president of the National Wildlife Federation, sees few alternatives to the present system of government command and regulation. He says that "the federal bureaucracy's red tape is sometimes frustrating, [but] the substitutes for ESA [Endangered Species Act] mandates would be persuasion and consultation."[31]

If we believe the environmental report cards issued by such groups as the National Wildlife Federation and Worldwatch Institute, increasing government regulation has coincided with steady environmental deterioration on most fronts. There may not be a cause and effect relationship, but it is clear that regulation alone is not the answer.

More and more people are beginning to see private property as the source of powerful new tools to protect natural resources. Every one of us who has ever bought property because of its trees or views has proved the point in the most obvious way. One observer of David Lucas's successful defeat of South Carolina coastal regulations said that in the future, if tree huggers want to save a tree, they will have to buy it.

Property ownership is a simple and well-tried environmental solution that continues to work. Except for parks and forests that have always been owned by government, almost all other protected lands have been purchased, but not always from willing sellers. The Nature Conservancy is the best-known purchaser among environmental groups. Since 1951, the conservancy has acquired some 5.5 million acres. In 1988's "Great North Country Land Auction," government and the Nature Conservancy competed successfully to buy 52,700 acres in northern Vermont and New Hampshire. After smaller bidders took 18,000 acres, the conservancy paid about $233 per acre for 7,700 acres in Vermont, while the New Hampshire State government won 45,000 acres of watershed north of its famous White Mountains for only $282 per acre. These lands were part of approximately 89,000 acres that a large developer had bought from Diamond International, a paper company. By selling the land, the paper company could reinvest in more profitable land or production facilities. The developer took a good profit on the lands it sold and walked away with several thousand acres whose cost had now been mainly paid for by profits from the sales.

The Nature Conservancy and other organizations also economize by buying only those ownership rights they really want. Ducks Unlimited, founded in 1937, has improved more than six million acres of habitat in several countries. In many cases, Ducks Unlimited has to buy only a conservation easement and not the property itself to protect the birds' wetlands.

Among the limited rights on which environmental groups should be able to bid are timber rights. Controversies over the spotted owl and other endangered species could often be resolved by opening the bidding on the timber in the embattled areas to environmental groups, although the government would have to eliminate a standard clause contained in its contracts when timber is sold. This clause requires the winning bidder to cut the timber within a certain period of time. Terry Anderson of the Political Economy Research Center admits that environmental groups might not be able to buy all the timber "because they do not have a blank check to write on. But neither does the American taxpayer."[32] However, to generate income to buy more lands or rights, the environmental groups could apply new, ecologically sound forestry techniques to harvest some of the timber they can afford.

Deseret Land and Livestock Corporation and International Paper are two companies that have proven how environmental protection and business can mix. Ecologically sound management of Deseret's 200,000-acre Wyoming ranch has allowed the company to increase both its cattle and wild elk herds while earning extra income from hunting fees. International Paper's holdings in southeastern states are larger than Yellowstone National Park. The company has set aside areas for wildlife, and hunting, camping, hiking, and fishing fees have made the program profitable.

These are examples of individuals and corporations responding to an incentive. Like my friend Stanley Selengut who owns the Maho Bay ecotourism facilities, they are making a profit by doing good. As anyone who has ever been in business knows, profits are nice, but they are twice as nice when you have also done something the world really needs. Unfortunately, instead of encouraging environmental preservation through self-interest, our society has increasingly created disincentives. Private efforts to use rather than consume natural resources such as forests have been hampered by what might be called unfair dumping by competitors. The competitors are state and national governments. Environmentalists have rightly protested the below-cost sale of timber from national forests, so why don't they protest the below-cost sale of recreation? Government habitually sells recreation rights (that is, entry fees) at far below actual market value. Sometimes it gives them away. This is very nice for the few poor people who use public lands, but it is a big discouragement for private landowners who might like to preserve their own property for the same purposes. Hunting rights to 48,000 acres of federal B. Everett Jordan Lake near my home are free. That is a powerful disincentive for local landowners to preserve wildlife habitat. Only eventual overcrowding of free government trout streams will encourage private landowners to nurture their own streams.

When it comes to selling rights to publicly owned resources, one thing government can do is level the playing field for environmental groups. This means eliminating subsidies for exploitative uses. It also means opening leases and sales of everything from water to oil, gas, and mining rights. Environmental groups in the United States have revenues that may top $1 billion. (In 1989, the Nature Conservancy alone had revenues of $109.6 million.) The huge amount these groups spend on lobbying and court cases could protect more resources if it were spent buying leases and options. Nothing would prevent them from turning around and subleasing oil, gas, mineral, and grazing rights under conditions they feel are environmentally sound.

Some land conservancies and private developers have already done this. They buy land with important natural resources, set aside the vital areas or write covenants to protect those resources, and then allow controlled development under their rules to pay their expenses. In Brazil, naturalists have proposed that chemical and pharmaceutical companies buy rights to set aside forests as biodiversity preserves where they could prospect for valuable chemical compounds that might lead to new fuels or drugs. In Siberia, some local governments have welcomed privatization as a way to stabilize reindeer production by assigning herds to private owners who will be more sensitive to overgrazing of tundra and forests than the Communist bureaucrats who tried to maximize short-term production.

Examples such as these encourage the idea that private ownership itself can be an environmental resource. But before this idea can realize its potential, government must cut some red tape. The Political Economy Research Center (PERC) says that salmon fisheries are a case in point. The PERC researchers Donald Leal and Terry Anderson point out that the dams and irrigation works that have starved the salmon streams are government projects that subsidize irrigation by as much as 90 percent. Allow owners to trade water rights on the open market, they say, and the salmon can be saved. They point to successful water purchases by the Nature Conservancy on Colorado's Gunnison River to maintain stream flows for the humpbacked chub.[33]

However, such purchases are rare in the West because water regulations require users to apply all their water to "beneficial" uses. Allowing water to flow naturally down a river does not qualify as beneficial. In fact, selling rights to leave water in the stream is a forfeit, endangering the seller's allotment and the buyer's lease. To bargain for in-stream water, fishing interests and environmentalists would have to deal with slow government bureaucracies.

The biologist Garrett Hardin coined one of the most repeated phrases in debates about the environment—*the tragedy of the commons*. It means that common ownership of a pasture or any other resource invites exploitation. Hardin proposed that humans inevitably act in their own self-interest (much like my landowning friend Professor X). If the profits from use outweigh any costs, each person will use as much as possible. Hardin used medieval England's pastures as an example, but other proof is everywhere: the oceans are overfished; peasants tear down public forests for fuel; farms and subdivisions exhaust underground water reservoirs; and industry fouls the atmosphere.

No place demonstrated Hardin's tragedy more horrifyingly than the world of communism. Outside every apartment door and beyond every factory fence was the commons, the people's property. It sounds nice, but it still smells awful. Go into almost any apartment house in the former Soviet Union. Residents often keep their individual apartments immaculate, but I've never seen an apartment house staircase that wouldn't be at home in a medieval dungeon—dark, filthy, garbage-strewn, and often smelling like an uncleaned bathroom. Few impoverished Third World countries have cities with dirtier air or rivers with more poisonous water. Perhaps if these once Communist societies had produced great wealth, all their common areas would have sparkled like the Moscow subway.

The former Soviet Union is an extreme case, but start to count the world's biggest environmental problems. Where are they? They are in the commons: air pollution, water pollution, depleted ocean fisheries, and overcrowded parks, national forests, and rangelands.

The 1992 Earth Summit in Rio de Janeiro demonstrated that environmentalists still have not got the message. The solutions proposed in Rio took two paths. Those most celebrated among environmentalists, such as the agreements on biodiversity and greenhouse gases, proposed environmental protection by command rather than incentive. Consciousness-raising, or education, was the second type of solution to emerge from Rio. If governments would convince their citizens of the seriousness of environmental problems, perhaps citizens would be willing to set aside self-interest and make sacrifices for the general good. Many ecologists and animal behaviorists find it odd that environmentalists have not yet discovered that throughout the biosphere, cooperation is sustained by self-interest.

Game theorists call it "tit for tat." It is a lesson that every salesperson and marketing department learned ages ago—happy customers keep coming back. If it pays to cooperate, people cooperate. Environmentalists recognize this on an abstract level, safe from the dirty details. That is why they are forever saying that a sound economy depends on a sound environment.

Free-market environmentalists start from the self-interest embodied in private ownership: assign private rights in the national and global commons areas, and property owners will have a personal economic reason to protect the areas. For example, with current technology, they say, we can even "fence" the atmosphere. Since all activity pollutes, the problem of clean air is not pollution, but overpollution. Once government determines how much pollution is acceptable, it can privatize by selling the right to emit a certain quantity of pollution. A company can then decide whether to install a smokestack scrubber to take sulfur out of coal or buy the right to put the sulfur in the air. Companies that choose to buy this right can purchase it from companies that choose to install scrubbers or use other means to reduce their pollution. Thus, the market achieves the same effect as laws limiting each plant's pollution or requiring a certain technology. However, each business has the freedom to choose the best economic solution for its circumstances.

In fact, Congress did amend the Clean Air Act to allow coal burners to buy and sell pollution rights. In March 1993, the Chicago Board of Trade auctioned off the rights to put 153,010 tons of pollutants in the air. Industry paid $21 million. Government protects these property rights the way it protects others—by acting as a policeman. Technology has made it possible to identify how much pollution is coming from a smokestack, and even miles away from the plant, tracers added to the burn process "brand" a company's emissions.

Through the 1970s and into the Reagan era, environmentalists condemned selling pollution rights as a license to pollute. Today, some environmentalists have endorsed the tradable rights approach to limiting pollution. However, the power to regulate is more appealing to them than the free-market aspect. Herman Daly, an environmental and economic expert at the World Bank's environmental program, has suggested salable rights to limit basic human freedoms:

> The bubble concept, which limits aggregate polluting emissions in a region and then allows market exchange to allocate the quota among alternative users, is one that I wholeheartedly support. With obvious

caveats about political acceptability, I would even suggest applying the same scheme to population control by setting up tradable reproduction rights corresponding to aggregate amount to replacement fertility.[34]

Although liberals sometimes unfairly accuse free-market environmentalists of being crassly materialistic and inhumane, Daly's proposal does suggest that in keeping the public good superior to individual rights, people might even sell their unborn children. Well, maybe other people's unborn children.

The free-market enthusiasts at PERC recognize that the market does not rule everything. When the Carter administration left the nation with 18 percent interest rates, holding on to mature private forest land did not make much sense; the value of forests grows slowly—3 to 7 percent a year is typical. So why didn't massive numbers of landowners sell their trees and put the money in the bank, utility stocks, or money-market funds, where it might have earned 10 to 20 percent a year? Some were simply ignorant. Others thought the price of timber would go up. But contrary to the environmentalists' dark view of human nature, both corporations and individuals looked beyond short-term economic gain. Some compared the personal value of beauty to the value of lumber and found that the trees were worth more if left uncut. Others wanted to survive as timber producers over the long term rather than taking the money and buying into a new business. However, at other times, money and its benefits outweigh all other reasons a private owner might have to protect a resource or an ecosystem full of resources. Families in besieged Sarajevo that were without electricity, gas, and oil had no compelling incentive not to cut down their orchard and shade trees for survival.

Where incentives and the free market won't work, the PERC senior associate Richard Stroup says that we must establish priorities, focus on habitats rather than species, and budget the money where the needs are. However, there must still be tit for tat as in the marketplace. If no one else will provide the monetary incentive or compensation for acquiring something that the public needs or demands, the government must. Taxing a few individuals for the sake of the many has always been a recipe for political trouble. It has already cost the environmental movement and the government enormous good will.

If the government does not take property without payment for the common defense in time of war, Stroup argues, why should protecting wetlands or endangered species be different? Paying for preservation may be expensive, but there may be no choice. The courts seem ready to implement Justice Holmes's judgment that when the degree of loss becomes severe, a taking has occurred. In any case, under our system of government, making enemies of landowners guarantees precious funds will end up in lawyers' pockets, and too often the environment will lose.

Fortunately, taxes do not have to be the sole source of compensation when regulations take property. One of the most talked about, least expensive, and yet seldom tried mechanisms for regulating land use is the transferable development right (TDR). In theory it is simple. A county or state assigns each and every acre of land a certain number of development rights—say ten. Then it zones. Some land is permitted one house per one hundred acres. Maybe it is a forest where the city wants a greenbelt. The forest has ninety-nine TDRs it cannot use. Meanwhile, an acre in the city is allowed one hundred apartments in a high-rise. Like a single acre of farmland, it has ten rights. It needs ninety more to build the whole project. Here the market takes over. The developer advertises for rights just as it might advertise for land. The forest owner can sell his unused rights to the developer. They negotiate the price just as they would negotiate for land. In the end, the forest owner has sold just one stick out of his bundle of property rights. He can still cut timber, lease hunting rights, and so on. The community has the benefit of a forest nearby. Except for setting up the system, taxpayers have not paid a nickel and government has not taken a landowner's investment or savings. Unfortunately, TDRs are seldom tried until development is well under way, and installing the system is very complex. Montgomery County, Maryland, adjacent to Washington, D.C., has one of the country's most sophisticated TDR systems.

Even simpler is straight barter. The federal government already owns between one-quarter and one-third of all the land in the United States. Using the new National Biological Survey as a guide, some of this land could be sold and the proceeds used to purchase conservation easements on ecologically valuable natural lands. Alternatively, as the San Francisco attorney Mark Pollot has suggested, the federal government could compensate landowners whose property it wishes to preserve simply by trading land with them. Both methods would be a kind of trading up for the environment and maybe an economic boon for the landowner. The trading-up policy actually could have achieved President Bush's goal of "no net loss of wetlands" and also a goal suggested by the late economist Warren Brookes: "no net loss of private property."[35]

People such as Ann Corcoran and Peggy Reigle believe that's the way it ought to be when private landowners are asked to bear the burden of regulations. They believe that when environmentalists recognize the rights of property owners, they will have reconnected with mainstream America. According to Corcoran, that will begin with "the first environmental group that works with landowners, instead of against them."[36]

CHAPTER 9

Technology to the Rescue

We may get to the point where the only way of saving the world will be for industrial civilization to collapse.

—Maurice Strong, organizer of the 1992 Earth Summit
in Rio de Janeiro

Suggesting that industry and technology will solve environmental problems sometimes evokes the same sort of response as suggesting that the Hell's Angels should chaperon the senior prom. Since the onset of the Industrial Revolution some three hundred years ago, nature lovers have focused hard on its destructive power. The poet William Blake penned the phrase still used to characterize industry—"dark Satanic Mills." For the novelist Sherwood Anderson, industry turned the rural citizens of Winesburg, Ohio, into greedy factory owners and robotlike workers. Thoreau retreated from industrialized New England to the woods. The bias against industry grew stronger, not weaker, as everything got better. Through guilt by association, science and technology became accessories to all of industry's crimes.

In preferring the simple life and its ultimate expression—wildness—environmentalists are exhibiting typical human behavior—trying to atone for all too human imperfection by worshipping an impossible ideal. Like so many of the guilt-ridden righteous, they also require sacrifices to atone for human sins. Too often these sacrifices are science, technology, industry, and the system of individual freedoms in whose shelter they survive. And they are just as willing to sacrifice other people's comfort and lives. Their energy conservation programs assume that poor people should pay for turning aside new technology in both coal and nuclear power. Their formulas for saving rain forests assume that the natives want to continue throwing spears, sleeping with cockroaches, fighting disease with herbal medicines, and dying at forty. Rather than assuring affordable disease-free chicken by irradiation, they assume that poor people prefer the more expensive Swedish cuts and free-range birds. Instead of promoting cheap flour, rice, and vegetables from large-scale farms, they assume that the world's poor would rather double-dig small garden plots and weed with hoes.

In 1977, Friends of the Earth, headed by the "archdruid" of environmentalism, David Brower, attacked the industrial notion of progress in its treatise *Progress As If Survival Mattered*, by Hugh Nash. On the cover, a railroad track forks, one line leading to skyscrapers in a cloud of black smog, the other to a grove of trees under a shining sun. That image sums up an increasing and deliberate polarization.

Few environmentalists stand up and oppose technology outright. They do not demonstrate in front of Congress and state legislatures, shouting "Down with technology." But with their own brand of conservatism, they have proposed endless ways of delaying and frustrating new technology. Vice President Al Gore, for instance, has proposed a review board for new technologies. It would operate like the council that former vice president Dan Quayle chaired to review the costs and dangers of new regulations.

The Food and Drug Administration's rules require millions of dollars in tests and years of experiments before it will approve a new drug, even if the drug has already been tested and approved abroad. America's nuclear power plants are the safest in the world, in part because environmentalists have insisted, but they are also the most expensive. In fact, they became so expensive that utilities quit building them and returned to coal. To the great joy of environmentalists, some utilities became converts to energy conservation programs based on demand-side management, or DSM. Utilities practicing DSM undertake a variety of public relations and advertising programs to encourage customers to cut demand. A common incentive is lower rates for people who install energy-saving technology. I'm all in favor of DSM if it means I don't have to live near a new power plant—nuclear or coal. And I'm willing to pay the costs even if my friends in the environmental movement say there are none. There are.

Energy conservation may be another expensive rejection of technology that pleases those who can afford it most but pushes the greatest burden on those who can afford it least. Energy conservation programs may be energy-efficient, but they are not always cost-efficient. The difference is clear if I propose that instead of your $300 washing machine, you should buy a $1,000 machine that uses less electricity and gets your clothes just as clean. You are operating at greater energy efficiency, but if you save only $5 a month on your electric bill, the extra $700 you spent won't be recouped before the machine dies.

Consumers do not cut back on electric use just because the power company asks them to be nice folks and help avoid the cost of a new plant. The principle of self-interest says that most people want something in return. Therefore, the power companies offer lower rates, rebates, and a variety of goodies that cost money, but you can bet your meter box that a power company is not going to offer something unless it makes money, too.

Electric companies like to advertise their conservation success because it makes them look good, even when they have raised rates to pay for that success. Why does conservation sometimes cost more than using new technology in new plants? First, many of the consumers who claim the rebates and lower rates would have adopted conservation measures anyway—especially those with a realistic pay-back period. Among these are turning back the thermostat and building well-insulated homes and additions. A study by Synergetic Resources Company found that 63 percent of the customers claiming benefits under DSM incentives would have invested in the conservation measures without being paid.[1]

One of the environmental heroes of energy conservation is the Rocky Mountain Institute's Amory Lovins. While he has proposed a number of interesting strategies, a recent study from MIT found that the cost of conserving a kilowatt hour is 500 percent higher than his estimate. The study's authors, Paul Joskow and Donald Marron, said that utilities often ignore the administrative costs of conservation programs. When they compared engineering estimates to actual statistical analyses of cost efficiency, they found that engineers often overestimated savings.[2]

Like energy technology, biotechnology has also been pinned down by hostile fire from environmentalists. Environmentalists give lip service to feeding the poor—growing more food and growing it more cheaply—but they opposed the technical means of doing so even before biotechnology came along. The Green Revolution, which began in the 1960s with new plant hybrids, is among the environmental movement's most frequently cited villains. However, environmentalists are unclear about what else could have saved millions from starvation. Production increases in only one of two ways: increase the amount of land cultivated or the yield per acre.

More than 80 percent of the increased food production in the recent past has come from increased yields. This was the Green Revolution. If it wasn't welcomed by environmentalists, the poor of Asia, Africa, and Latin America who ate more and worked less in the fields were grateful. Environmentalists might want to consider how much more land—much of it in pristine areas—would have been cleared to feed the world if the Green Revolution hadn't happened. More important, if we don't use biotechnology to maintain yields, fight pests, and adapt a larger variety of plants for difficult, small plots, are environmentalists willing to tell people to clear and plow more forest land or else starve?

While environmentalists preach against technology and try to deny it to poorer people, they are heavy users of technology once it arrives. They give little thought to the chemicals, energy, and technical processes used to produce Vibram soles for hiking shoes, Gore-Tex rain gear, freeze-dried camp food, superstrong rock-climbing ropes, Volvos, Kevlar canoes, and magnesium-alloy pack frames. They live in homes well insulated by spun fiberglass and double-pane reflective windows. Kitchen counters are made of plastic laminates, cabinets of plywood or chipboard held together with high-tech glues. The list is long and growing.

I am reminded of a meeting of nature writers at Williams College in 1990. Everyone talked about consuming less and living more simply. So I took a quick poll. How many at the meeting lived in a home smaller than the American median size of some 1,700 square feet? How many of our cars parked outside the old Rockefeller mansion conference center got better mileage than the average fleet target? No homes qualified and only two of more than a dozen cars did. Since none of us had a large family, we were actually consuming more per person than other people in the world with similar possessions. The per capita hypocrisy was shameful. If even this group of distinguished and widely published environmentalists cannot live simply, then cutting back on technology and consumption is not the answer. In a world of six billion people headed fast for twelve billion, more technology, science, and industry are the *answers*, just as they were for child laborers, ever pregnant mothers, and diseased infants of the 1800s.

CHOOSING TECHNOLOGY

Just as a quick poll shows that a simpler life and salvation through "appropriate technology" such as pedal power are unrealistic, a quick look at how industry is using technology shows that real solutions are blossoming all around us. They are blossoming partly because the moral indignation aroused by environmentalists has created a market and partly in spite of that indignation.

In the fall of 1992, the American Re-Insurance Company sponsored a meeting of thirty-one companies with annual sales ranging from $1 billion to $40 billion. On an environmental survey, they listed waste reduction as their single activity with the greatest social benefit. The two greatest benefits to the companies themselves were complying with environmental regulations and waste reduction. Participants strongly felt that one of the greatest obstacles to more effective cleanup efforts was the "command and control" micromanagement that limits the technologies industries can use or develop for environmental improvement.

American companies estimate that their environmental expenditures for the 1990s will top $1.2 trillion. It is easy to see why they do not want a Washington bureaucracy telling them what technology to use. The incredible pollution of Soviet industries was in large part caused by a central bureaucracy dictating not only what to produce but also how to produce it and what technology should protect the environment.

Environmentalists, regulators, and polluters often work against innovations that would be both cost-efficient and environmentally sound. Worries about acid rain brought power plants under the gun in the 1990 Clean Air Act. Power plants were producing 25 percent of the nitrogen oxides that turned into nitric acid. When EPA announced it would fulfill Congress's mandate by choosing one of two available technologies to reduce nitrogen oxides, environmentalists and the producers of catalytic products pushed for the catalytic process, while Combustion Engineering and utilities pushed for low nitrogen oxide burners. The burners removed only 30 percent of the offending gases, but were relatively cheap at $200 per ton of pollutants. The catalytic process removed 90 percent but cost $2,500 a ton. EPA chose the cheap route, blessing Combustion Engineering with a captive market and making it very risky for anyone to develop new products. Almost immediately, a third product emerged that removed 60 percent of the gases at only $414 a ton. Produced after years of expensive development by Nalco Fuel Tech, the product was twice the price of the burners, but twice as effective. And it did not require replacing the entire old burner. While EPA and the U.S. Office of Management and Budget haggled over how to change the rules, the company's orders fell off. It was losing money. In 1993, Nalco finally began to get new orders, but the delays have been expensive for both the economy and the environment. If environmentalists, industry, bureaucrats, and polluters would agree to set air standards rather than specify technologies, creative people would have more hope of getting rewarded for solving problems efficiently.

The story of ozone-destroying chemicals demonstrates the success of giving the marketplace its freedom. Whether the common refrigerants, chlorofluorocarbons (CFCs), really are destroying the ozone layer is not the point here. They have been outlawed, and industry has responded with a variety of alternatives. They range from using other high-tech gases to cooling by sound. At the Naval Postgraduate School in California and at Los Alamos National Laboratory, once dedicated to atomic weapons development, researchers have already tested a "thermoacoustic cooler." The device is complex but adaptable to large and small uses. In simple terms, a loudspeaker in a tube produces an extremely loud note (which is hardly audible outside the tube). Vibrating molecules cross a metal plate that conducts their heat away from the refrigerator. Gregory Swift, one of the refrigerator's developers, told the *New York Times,* "With $20 million and a few years, we could make thermoacoustic refrigeration a commercial reality."[3]

While pundits everywhere cry out against the decline of social values, many American businesses are responding to a rising tide of consumer morality. Economists often say that consumers vote their values with their dollars. For more and more consumers, those votes are guided by moral concerns about the environment. Recent research by the Western Wood Products Association (WWPA) revealed that by a two-to-one margin, "adults would sacrifice economic growth to keep the environment clean."[4] WWPA also says that some 30 percent of its customers now worry that by buying wood products they may be harming the environment.

Perceptive business experts have asked industry for products that respond to these concerns. Business has the ultimate responsibility for turning science into technology, and technology into new products and services. Industry has begun to answer environmental demands. Motives range from pure profit to love of nature. Environmentalists also have mixed motives—personal pleasure, scientific study, revenge, and political agendas. Let us admit the general imperfection of human motivation and move on to what is happening.

The response is moving quickly on three fronts:

- More efficient production: making more things with less material
- Finding new resources: substituting new products for old ones
- Benevolent technology: inventing new, kinder, gentler technologies

MORE EFFICIENT PRODUCTION

Ever since the beginning of the Industrial Revolution, the profit motive has led producers to try to make more with the same amount of raw materials or less. The debate over clear-cutting spectacular, large trees has consumed the media's attention and conveyed the impression that America's forests everywhere are disappearing. They are, in fact, expanding. The reasons are simple and all result from more complex technology:

1) Log cabins and post and beam yielded to the two-by-four and the frame house using smaller amounts of more plentiful second-growth softwood lumber.
2) Producers of wood products have used technology to make more with less.
3) The good old days were bad days for forests. As farmers and travelers converted from horses to tractors and cars, from wood heat to coal, gas, oil, and electricity, trees came back.

I can see the results all over the two thousand acres of woods near my house in North Carolina. I can locate at least ten huge piles of sawdust, pine bark, and rotting slabs within a half mile of my house. They may eventually renourish the earth, but for fifty years or more, nothing but a few vines has grown through the piles. The heavy, thick circular saws that whined through the woods and at the mills destroyed as much as 60 percent of the logs they cut. Tough steel band saws are five times thinner, and sawmills no longer throw away slabs and sawdust. Timm Locke of Western Wood Works says that one hundred years ago, probably 70 percent of incoming logs left a mill as usable product. Today it is close to 100 percent. Most western mills shun oil or electric energy created by nuclear fuel or coal, burning sawdust and scrap instead.

Fly over New England, the mid-Atlantic, or the Southeast and you can see clearly the old field lines that bound hundreds of thousands, if not millions, of acres of second-growth forest where once little but grass or crops grew. (Okay, tree farms aren't real forests, but they do shelter wildlife and lock up carbon dioxide.) Technology and industry made this greener world possible. Almost by accident.

FINDING NEW RESOURCES

When President Clinton and Vice President Gore promise that environmental technology will be an economic boon, they focus on technology to clean up wastes, scrub smokestacks, and filter water. This remedial technology is only a sideshow compared to what is already happening. American business leads the world in creating products that give us more for less—less money and less environmental damage.

The idea of profitably substituting one product for another has been around since stone arrowheads replaced wood, grains replaced wild animal protein, and coal replaced the dwindling supplies of oak needed by glass, iron, and mortar makers before the Industrial Revolution. (Which ought to remind us that the Industrial Revolution did not denude medieval forests.) However, today environmental considerations play a big role in choosing substitutes. This is especially true when market mechanisms include environmental costs.

Almost any modern house uses dozens of new products that are replacing not only old-growth timber but other kinds of wood as well. Because of today's new environmental concerns and the changing nature of forests, the new technologies are paying profits—a necessary result if they are to last and grow. The Minnesota Trus Joist plant, which uses shredded aspen to replace solid lumber, generated some $25 million in first-year sales from a $75 million plant, and it expected to double that profit in 1993.

Even the pencils used by carpenters and schoolchildren are being made from a wood substitute. In New Jersey, Faber-Castell introduced the American EcoWriter, a nontoxic pencil made from old newspaper and cardboard that is almost indistinguishable from the familiar wooden pencil.

A new process developed at the University of Washington may remove as much as 30 percent of the wood pulp used to make the paper we write on. While most paper contains clay filler between the fibers, Graham Allan, an engineer specializing in wood products, has found a way to introduce a filler inside the cells of the wood. The difference in the standard process that uses lots of pulp and Allan's process is that he inserts a filler into the pulp fiber before it dries. This stops shrinkage and requires less wood fiber.

The balsa wood forests that are being depleted for everything from model airplanes to insulating supertankers and muffling the noise in train cars may be saved by a new seaweed product. At California's Lawrence Livermore National Laboratory, Robert Morrison has invented biodegradable SEAgel made from kelp. It can be manufactured in varying densities to replace drug capsules, refrigerator insulation, and packaging foam. SEAgel's lightest form is lighter than air, and a thick piece can rest on a soap bubble.

Not everyone is moving away from wood. Twiggy models at the next fashion show will be wearing more biodegradable wood instead of rayon or acrylic. Tencel, the new wood-based fabric created by Courtauld Fibres, is a good stand-in for wool whose sheep require cleared pastures and for cotton whose cultivation is chemical intensive.

Just when landfill operators are wondering if plastic packing materials will overflow the dump, technology offers several solutions. Everyone who curses the Styrofoam "peanuts" that spill out of packages will be glad to know that new high-tech peanuts can be dumped in the kitchen sink or thrown out in the rain. They will disappear within seconds of getting wet. They are made from cornstarch by American Excelsior in Arlington, Texas.

Customers won't be happy to discover that the new packing costs twice as much as Styrofoam. Forestry Suppliers says that packing costs for an average half-cubic-foot box have risen from twenty-five to fifty cents, but the company has made the switch because its environmentally conscious customers hate Styrofoam. Sometimes the increased product costs are offset by the public relations value, the reason that such high-profile companies as Kodak have begun using the new peanuts. American Excelsior, which still makes shredded wooden packing from aspens as well as styrene peanuts, says that it expects soon to have the price of the starch product down to within 25 percent of the plastic one.

Once again, high technology has produced big environmental benefits. In the case of the peanuts, manufacturers transferred technology from high-tech cereal to packing materials. You may not know that starch helps keep your cereal from going limp in cold milk, but it does. As a starch maker was trying to improve the technique, a researcher suggested using a starch product for packaging. Salesmen sometimes eat the peanuts to demonstrate their environmental kindness, but to avoid bugs and rodents, all the flavor and nutrition have been removed.

BENEVOLENT TECHNOLOGY

In our health-conscious society, and with land prices soaring, few industries have sought new technology more intensively than the garbage handlers. No honest environmentalist can badmouth high technology and promote recycling at the same time. One of the greatest obstacles to recycling has been the cost of present technology. As most homeowners know, it is cheaper to buy a new piece of lumber than to hire someone to tear down a building and sort, denail, and trim old lumber. The same applies to newspapers, tires, plastic bottles, and metal cans. The best recycling programs divert only a fraction of our garbage and often cost four times more than regular collection. Subsidizing recycling is yet another new tax, and to favor recycling by taxing virgin materials hurts the poorest consumers most.

Any wide-scale, cost-effective recycling effort requires complex and innovative technology. In Montreal, Tigertec Environmental Services has been working twenty years to develop a system that takes bags of unsorted garbage. After separating the metals, it shreds, heats, and composts everything else in a short, forty-eight-hour trip through long, rotating "bio-barrels." Plastics and fabrics that have not degraded are reduced pyrolytically (that is, using heat without oxygen) to pyrocarbon for metallurgy, gas, and tar. Tigertec says that its largest plant has recycled 11 million cubic meters of solid wastes and produced 1.7 million tons of compost, saving some 655 acres of land from becoming a garbage dump.

Last year in arctic Russia, while I was traveling the Kolyma Trace, a road that destroys trucks as easily as it once destroyed political prisoners of the gulags, I saw little roadside garbage, except for a few hundred thousand old tires. My host said local environmentalists had often wished that someone would be inspired to find a use for them. American landfill operators and health officials have the same wish for the 200 million plus tires that die here every year. To make tire recycling pay requires a high-tech total-use approach. In August 1992, the Florida inventor Charles Ledford received the state's first permit to operate a commercial tire pyrolysis plant where heat and chemistry reduce old tires to oil, carbon black for printing, rubber crumbs, and methane. In the airtight recycling process, the plant uses its methane production to run itself. Ironically, liners for landfills were one of the first products made from Ledford's rubber crumbs.

The pyrolysis process is now being developed by a publicly traded corporation, ECO2. The company estimates that from one ton of tires it can recover 110 gallons of fuel, 500 pounds of carbon black, 150 pounds of steel, and 4,000 cubic feet of gas.

The computer has been accused of generating more paper than it has saved, but the tide is changing. Giant worldwide computer networks such as Internet daily transfer quantities of information that would require multiple editions of the *New York Times*. A lawyer friend of mine set up office with a whole wall of law books contained on a single compact disk for her computer. Every phone book in the United States is now available on a few ounces of disk. Checkfree Corporation, founded in 1981, employs only 250 people, but it handles $3 billion a year in paperless checks that can be written by single homeowners or large corporations.

The present trend in storing more information in less space could take a great leap before the end of the century. Scanning probe microscopes (SPMs) can now sense the presence of individual atoms and even move them about. If atoms can be moved about to become the digital signals by which a computer stores information, researchers estimate that the entire Library of Congress could be stored on an eight-inch silicon wafer.

All over America, writers like myself, not to mention people in universities and corporate offices, no longer print everything on paper, pack it in an envelope, and send it by internal combustion engine to editors and colleagues.

The fastest and most widespread environmental benefits of technology are beginning to pour forth from an industry that terrifies environmentalists. For many environmentalists, biotechnology is the ultimate in human arrogance. Despite a lot of publicity about oil-eating bugs, these new critters have not been bred well enough yet to clean up an *Exxon Valdez* spill or the oil fields of Kuwait. However, new bacteria have gone to work cleaning grease out of the Los Angeles sewer system's old pipes, saving millions in replacement costs. Environmental Biotech, which produces and "trains" the new bacteria, has also applied them to the palm oil residues that were killing Malaysia's food-rich estuaries. In Australia, the company is developing bacteria to save shrimp, fish, and krill from sugar refinery wastes.

In agriculture, biotechnology has been cursed for creating crops that allow farmers to use more herbicides, but it is also on the verge of developing pesticides that are more discriminating about which of nature's creations they attack. Ciba-Geigy is among the much maligned multinationals betting heavily on cell research to formulate insecticides that are tailored to the cellular chemistry of only harmful insects. Such developments as these could replace the scattershot pesticides commonly used.

Even the deadly chlorinated organic chemicals that persist in the environment may soon be destroyed by biologically engineered microorganisms. At the New Jersey Institute of Technology in Newark, Piero Armenante is working with a white-rot fungus discovered in Siberia. He expects to enlist it in a process that will break down chlorophenols that pollute many urban waste streams.

Deadly pesticides are relatively easy to produce. An alternative, safer treatment for termites is high-tech. In California, Tallon Termite and Pest Control uses liquid nitrogen to "freeze their little buns off."

By far, the greatest benefits to both the economy and the environment will come from alternative energy technologies and from technologies that save energy. The first generation of solar electric cells and heaters for homes and water systems blossomed in the 1970s after Saudi Arabia led a shutdown of oil supplies to America. They were often energy-efficient, but seldom cost-effective. They replaced coal and nuclear electricity, but purchase and operating costs made each kilowatt hour more expensive. They were simply not sufficiently high-tech. Miniaturization, computerization, and new materials will soon make alternative energy systems competitive. The cost of solar cells for generating electricity has been cut in half several times over the past twenty years, and researchers agree that if it happens once more, the cells will begin competing with utility-generated power.

Solar-heated homes once required expensive architectural additions—roof panels, basement storage systems, extra interior masonry walls, banks of chemical storage cylinders and space for them. Today, industry has come to the aid of architects with new insulating products and intelligent windows that adjust their reflecting qualities to keep heat in or out. Any homeowner can buy a computerized thermostat for less than fifty dollars that will save its price in less than a year by controlling heating and cooling when no one is home.

But lest someone takes this as an endorsement of solar power subsidies, I should note that heavy subsidies for specific technologies always discourage innovations that might be even more promising—biomass energy, new chemical fuel cells, or hydrogen, for instance.

"Don't make mistakes you can't reverse" is one of the environmental movement's minor but important axioms. Or as the father of modern wild life biology Aldo Leopold warned, a smart tinkerer saves all the parts. Thanks to technology, we can now reverse mistakes that were once thought permanent, and we do not always need to find missing parts to recreate something we have lost. Largely due to science and technology, environmental restoration will soon become the positive side of environmental studies.

Despite *Jurassic Park*'s bad publicity for revitalizing ancient genetic materials, we may one day recover or create reproductions of extinct plants and animals. Perhaps they won't be dinosaurs, but genetic material from frozen Siberian mammoths has been distributed to a number of laboratories, and it is conceivable that it could one day be combined with material in modern elephants. We have already created frozen germ plasm from plants for future combinations that might save crops from unexpected plagues.

The analysis of genetic materials has made possible the conservation of rare stands of Central American trees. Researchers at North Carolina State

University, working with Latin American colleagues, have been identifying genetic varieties that are in danger of extinction in their home territories. Successful stands have been established in safe areas of other countries. One day they may be used to restore Latin America's highland forests.

In Russia's far northeast, where gold miners have torn up hundreds of thousands of acres of valleys, a small group of scientists using bulldozers to implement the designs from computer models has shown that even these fragile permafrost lands can be restored. In some cases, the scientists have managed to restore them so that permafrost lies deeper beneath the surface and a new soil mix creates a microclimate where local citizens can grow potatoes, once imported from European Russia or regions a thousand miles south near the Chinese border.

When environmentalists predict irrevocable losses if old-growth forests are clear-cut, they are shortchanging both technology and nature. To say that such losses are total and irreversible ignores that nature restored old growth after glaciers denuded the landscape more completely than bulldozers ever could. With restoration science and technology helping, nature's processes can be accelerated.

The same claims of irreversible loss are made for wetlands. While the environmental community continues to talk exclusively about the loss of wetlands, no one notices the progress being made in creating and restoring them. For years, the U.S. Corps of Engineers has destroyed wetlands, but it has also used dredged materials to build new marshes. In Israel, where planners admitted the mistake of converting a peat bog to farmland, technicians are restoring two thousand acres to swamps and lakes. They hope to recover the loss of some $15 million in farm revenues from increased tourism. On New York's Long Island, a developer has earned the right to develop some artificially created high ground by recreating five acres of marsh lost many years ago. The cost of $10,000 an acre will be compensated by profits from development. In Maryland, Ed Garbisch has been pioneering efforts to develop new wetlands that look entirely natural to most people. His company, Environmental Concern, begun in the early 1970s, has completed more than 350 projects. The secret to wetlands creation is the right combination of soils, plants, animals, and constant care.

The nation is nowhere near restoring its former wetlands, but we have restored several hundred thousand acres since the mid-1980s, and we are quickly developing the technology and science to create more.

Reforestation has been a common restoration process for centuries, but only in our time has it progressed much beyond a kind of tree farming or simple laissez-faire regrowth. In Costa Rica, Dan Janzen has become world famous for his careful and relatively rapid recreation of a tropical dry forest on 203,000 abused acres of Guanacaste National Park. By studying the animals that disperse seeds and the interrelations among plants and insects, Janzen hopes to create in 150 years what would have taken nature 500 years. Even cows help by keeping down grasses that compete with woody plants for soil nutrients.

Restoration science success stories are multiplying rapidly around the world, but not in environmental magazines and books.

NO PERFECT SOLUTIONS

The ABCs of Waste Disposal
NIMBY . . . Not In My Back Yard
NIMFYE . . . Not In My Front Yard Either
PIITBY . . . Put It In Their Back Yard
NIMEY . . . Not In My Election Year
NIMTOO . . . Not In My Term Of Office
LULU . . . Locally Unavailable Land Use
NOPE . . . Not On Planet Earth

—A poster seen at EPA

The list of new ideas and products that will allow us to get more from nature with less damage could go on for volumes and is growing every day. If it seems that many of the products I have listed are inconsequential, that's exactly the point—environmentally friendly technology has reached every part of our lives, from packaging to gardening.

As environmentalists constantly remind us, the list of possible dangers grows too. For every action there is a reaction, for every possible solution a possible problem. The world is not divided into appropriate and inappropriate technologies. From organic gardening to biotechnology, every action requires a trade-off. To produce the cornstarch packing peanuts, someone has to grow more corn and use more fertilizer and diesel fuel. To make SEAgel means harvesting kelp where sea otters feed and fish breed. Lumber substitutes often require high-tech chemical glues. We can save a lot of trees by replacing wood house framing with steel, but midwifing a steel stud takes up to nine times the energy used for wood, and steel framing requires 10 percent more insulation.

Environmentalists who would like to stop the conversion of forests to pasture generally oppose biotechnology that can grow more meat or make more milk from less land. Bovine Growth Hormone, for instance, increases milk production by up to 25 percent. On an average dairy farm with one hundred cows, it could liberate twenty-five to fifty acres of pasture. The treated cows do get more mastitis and need more antibiotics, whose residues could affect sensitive consumers, but lower milk prices would have a much more powerful health effect on poverty-sensitive poor children. On the other hand, low prices would hit the small farmer hardest.

Encouraging the use of new products in houses is fine, but many standards that environmentalists want to write into building codes often make housing less affordable. Other requirements make housing more attractive, a lure for consumers to buy something environmentalists should abhor—the energy-inefficient, material consuming, free-standing, single-family house.

The determined antitechnology and antibusiness crusade of environmentalists has led them to drown their own solutions in regulation and red tape. In 1991, when the California legislature passed a law requiring increasing portions of recycled fiber in newspapers, MacMillan Bloedel proposed a $1.5 billion recycling plant for West Sacramento. Every day the plant would have consumed 2,700 tons of old newspapers, magazines, and other paper scrap. "MacBlo" (the company's nickname) is a certified villain among environmentalists, one of the "most wanted" for its alleged pillage of Canadian Pacific old growth. The Sierra Club leaped to oppose the plant's permit, charging the company had polluted rivers in Canada. In October 1992, the Sacramento Valley Toxics Campaign sued the city, saying its environmental impact statement did not reveal the full impact of the plant. And when MacMillan Bloedel proposed generating the plant's electricity from its own steam, state regulators said that any generator making more than fifty megawatts had to submit to the elaborate permit procedures of the California Energy Commission.

Fighting on several fronts to do the recycling mandated by environmentalists, MacBlo dropped the power generation plans. It cut the plant's size and capacity by a third. It agreed to offset any air pollution by financing antismog technology for other area plants. And it agreed to pay the legal costs of the Toxics Campaign's lawsuit. After two years and more than $3 million in legal and consultant fees, MacMillan had not recycled one newspaper, or even broken ground. Nor had it yet begun jumping through the hoops for federal and state water discharge permits.

Perhaps Mother Teresa, Buddha, and St. Francis were far enough removed from the temptations of this world to unhesitatingly choose lives of poverty and simplicity. For the rest of us, such choices are indeed otherworldly. Those who wish to remain pure should heed Hamlet's advice and "get thee to a nunnery." In the real world, whatever our rhetoric and ideals, the choices we are willing to consider are far from pure.

Almost every day in every way, the rest of us demonstrate that we want technology to intervene between us and the awesome powers of nature. Whether we admit it or not, we all want and gladly pay for dominion "over the fish of the sea and over the birds of the air and over every living thing that moves upon the earth." The proof is in our breakfast cereals, computers, cars, homes, vacation travel, Gore-Tex hiking shoes, microwaves, maglev trains, electronic mail systems, home insulation, painless dentistry, reconstructive surgery, prostheses, glossy environmental magazines . . .

The critical choice at the end of this century is not between simple living and the apocalypse. In a world of six billion people, the only choice is among forms of high-tech, sophisticated dominion. These choices are being offered in overflowing abundance by the same Western business and industrial psyche that stands accused by everyone from Gore to Earth First of turning us against nature. Yes, the business that converts science and technology into industry is motivated by self-interest and often greed. But in a system that rewards these motives only if they fulfill customer demand, the real burden is on the customer.

I have described a few simple examples of how business can respond to environmental demands. The burden of matching industry to environment begins with consumer demands. Then it requires honest response and labeling by business. If environmentalists want to lead an effort to educate consumers, their first job is to recognize the complexity of our choices. Following false gurus of the simple life and animal rights is a luxury of intellectual and affluent Romantics.

Environmentalists often argue that if everyone in the world achieved the same standard of living as Americans and Europeans, we would exhaust our natural resources. Like most such gloomy forecasts, this denies human beings their most distinguishing trait—creativity. No society has applied that talent so well to the practical side of life—food, clothing, and shelter—as the free-market democratic societies born of Western culture. It is no accident that the greatest human suffering is outside this tradition, and that the most unrestrained materialism and industrial destruction occurred in Communist countries. And it is no accident that the environmental movement and the science of ecology developed in industrialized societies where people recognize that pursuing self-interest demands both free markets and free speech.

A few months before thousands of environmentalists met for the 1992 Earth Summit, a distinguished group of fifty scientists, sociologists, medical specialists, and ethics experts met in Heidelberg to discuss the impact of synthetic chemicals on the environment and human health. They intended their exploration to be a model of informed discussion whose results could set a standard against which to measure the rhetoric of the forthcoming Earth Summit.

The conference ended with the Heidelberg Appeal, directed to the presidents and prime ministers who would attend the Rio conference. It was eventually signed by more than eight hundred well-known scientists, including sixty-two Nobel Prize winners. The statement includes the following call to consider real-world conditions:

> We contend that a Natural State, sometimes idealized by movements with a tendency to look toward the past, does not exist and has probably never existed since man's first appearance in the biosphere, insofar as humanity has always progressed by increasingly harnessing Nature to its needs and not the reverse.
>
> We stress that many essential human activities are carried out either by manipulating hazardous substances or in their proximity, and that progress and development have always involved increasing control over hostile forces, to the benefit of mankind.[5]

The most deadly blow we can strike at the future of our environment would be to limit the capacity of business to respond to our demands by chaining it with impossible regulations and crippling taxes in the name of nostalgia for a golden age of simple harmony. Instead, we must expand our understanding to include the creative potential of our society. If Thoreau's own creative potential had not been so biased by his dislike of humankind, he might have realized the truth with which this book began, that in wildness is not the preservation of the world, but in civilization is the preservation of wildness and the rest of nature.

CHAPTER 10

This Is Not the End, But the Beginning

Humans, large terrestrial metazoans, fired by energy from microbial symbionts lodged in their cells, instructed by tapes of nucleic acid stretching back to the earliest live membranes, informed by neurons essentially the same as all the other neurons on earth, sharing structures with mastodons and lichens, living off the sun, are now in charge, running the place, for better or worse.

—Lewis Thomas, *The Lives of a Cell*

Environmentalists have been trying to kill the messenger bearing good news. The good news is that we have indeed entered a critical period in human and natural history, and we are superbly equipped to have a great time. The weather ahead is not just illuminated by a few "rays of hope," as the environmentalists Paul and Anne Ehrlich write.[1] There is a vast cornucopia of hope, a shining mother lode, not of metals or oil or timber, but of human knowledge and imagination. But the doomsayers have flooded the bookstores, magazine racks, and talk shows with their fears and warnings, almost suffocating the good news.

People who are frightened of change always magnify its potential for destruction until its potential for good is eclipsed. In his book, Vice President Gore overwhelms his hopes for technology with his fears about human nature. His one great push for technology is for the much needed electronic "information highway," a vast network of words and data.

Many people in the environmental movement who believe they are the vanguard of a new society are actually fleeing backward to old ideas of order. Yes, we have problems, but we are not sick unto death. The overall prescription coming out of the environmental movement is not much more advanced than leeches, bleeding, or purging. Yet even this reaching backward indicates that the world has begun something as transforming as a second Renaissance.

The Renaissance that followed the medieval era also reached backward, through Muslim scholars to the ancient Greeks and Romans. Renaissance scholars recovered the learning of Aristotle and Pythagoras, but it was the innovators Galileo, Luther, da Vinci, Bacon, and a new urban middle class that transformed Europe's view of nature and what it means to be human. Reaching backward for past values and inspiration is more symptom than cause. It is a sign that new choices are necessary and available. As Wes Jackson, a Kansas agronomist, said:

> I don't think that we will have discovered America until we have some sense of the feelings of the natives that walked over this place. What we are about is the discovery of what we have undone, and then we can begin to talk about what to do.[2]

In his work, Jackson is trying to coax native prairie plants to sustain civilization the way they sustained buffalo herds. When he talks about discovering what we have undone, he could be either a Romantic or a man of the second Renaissance, or both. If he means we must discover some primitive Eden or paradise, he is indulging in a nostalgic condemnation of the present. If he means that through science and technology we can now see in a new way the natural world we have exploited so clumsily, he is talking about the human endeavors that will characterize the second Renaissance's rediscovery of the earth.

A renaissance often begins with people caught by change or people needing change who look to the past for answers. The past, after all, is much better known than the future. The past is like grandmother's house—a place of refuge, simpler living, and temporary relief.

The Pilgrims who sailed for America often spoke and sang of "the promised land," an idea as old as the Middle Eastern cultures from which Judaism, Islam, and Christianity emerged. In the nineteenth century, French, German, and American communes tried to imitate early Christians or primitive societies. In the 1970s, young people thinking they were the vanguard of the peaceful, natural Age of Aquarius formed communes. Many called themselves "tribes" and adopted animal totems and primitive clothes, or simply shed their clothes to confront middle-class America with the human birthday suit.

On the right, the Moral Majority, the Sagebrush Rebellion, the Posse Comitatus, the Reagan Revolution—all attempted to recapture "traditional American values." Even the new ethnocentric Native American movements, tired of being nobodies in American society, think they will be happier if they can only recapture what their ancestors were. And Communists, whose public relations experts loved to call them "the vanguard of the people," were in many ways a return to feudal paternalism and the authoritarian church. All these efforts we might call "backward change."

The "forward change" of a true renaissance often emerges out of the sum total of backward change. Because forward change is a different direction than most movements want to take, it is seldom recognized until it is well under way. Even then, many people on the left and the right try to kill the messengers who bring the news. Except when in real danger, most people prefer what they know to what they must learn.

The sudden, intense debate about values is a sign that we have entered a second Renaissance, an attempt to rethink our place in the world and to redefine ourselves and our purpose, and to rediscover a higher will. In this pursuit, people who often despise each other have the same motivation. Each group has set out on a voyage of discovery as surely as the people who crossed the Bering land bridge at the end of the Ice Age and as surely as Marco Polo, Columbus, and Magellan sought new lands. Despite different directions, each group is part of a common quest. The word *renaissance* means a rebirth, of course. The Renaissance of the late medieval era was a resurrection of classic Greek and Roman culture. The very idea of nature was also reborn in the human mind, just as it is being reborn yet again in our time. A renaissance depends on many changes, and a few good futurists have covered the wide field quite well. I will limit my observations to our new understanding of nature.

I am talking about nature as we know it. This is a nature we create out of our limited perceptions, intuition, and emotions. Beyond that, out there where we cannot reach with either our senses or our technology, a broader nature exists. We can guess at it and imagine it, but we will never know it. The nature we know is nature filtered through our limited selves. Yet because we have minds to contemplate what we see, hear, touch, smell, and taste, what we understand nature to be can change radically as it did in the classical world, again in the Renaissance, and now at the end of the twentieth century. The human animal's greatest ages of achievement have always begun with enormous energy poured into rediscovering the environment. Discovery is not a one-time, unrepeatable event. In truth, every place in North and South America has been discovered at least twice.

The first discovery was prehistoric. Even the great deserts and rain forests that were not hospitable to the European lifestyle were settled at least for a while during the past 50,000 years by the diverse peoples we now lump together as Native Americans.

The second wave of discovery begun by Europeans in the sixteenth century scoured the New World for fertile lands, mines, timber sources, new foods, ports, and the unbaptized. The first three decades of the twentieth century brought this second phase of discovery to a sudden end as it unleashed its trains, boats, cars, and planes. From Machu Picchu and Mato Grosso to Point Barrow and Baffin Island, the New World had been visited a second time.

The Europeans, who wanted different things than the people they found and who had a more sophisticated technology for using resources to fulfill their desires, settled and exploited many areas that the first discoverers had passed up, although places such as the Amazon valley and the arctic tundra attracted little attention from either wave. The people who went to the most difficult places in the first wave of Stone Age explorers were often quickly lost to the rest of the world. The European explorers took a look around these places and went elsewhere. At best, they left behind a few traders or missionaries.

Except where the natural resources offered a great economic lure, the second wave of discoverers concentrated on the most comfortably and practically habitable regions of the New World. Even now, in the midst of an awesome population explosion, the human animal remains choosy about the places it calls home. The numerical explosion is in fact a rash of demographic implosions as more and more rural people flock to great cities, recognizing more promise in technology than in the lands that satisfied the needs of their parents and ancestors. People are coming to these cities as they once came to the cities of Europe during the Renaissance. And once again, as the newcomers arrive, others leave. Some flee city life. Some go to satisfy their curiosity, some for personal ambition. Some go seeking knowledge, others, privacy or pleasure. They are all setting out to be part of a third adventure in discovery.

The Ice Age migrations and the Renaissance voyages were the two greatest waves of discovery that humanity has ever undertaken. The third great wave of discovery has begun, and it is very different from the spread of Stone Age people or the intellectual and physical travels that brought European culture to every corner of the world.

THE END OF THE SECOND WAVE OF
DISCOVERY

The Renaissance that brought us modern science, the Industrial Revolution, and unprecedented improvements in the human condition began in environmental disaster in the 1300s with the Black Death and ends in the twentieth century as we accept our role as the caretakers and, yes, the managers of all life on earth. To arrive here required a shattering of the confident world created by the first Renaissance. The necessary end of the Renaissance vision of nature and our place in it was signaled by several events that vanquished Western self-confidence. These events were two world wars, the Great Depression, the spread of communism, the Vietnam War, and environmental threats of a global scope. Each of these events, in its own way, cast doubt on the greatest material consequence of the Renaissance—the Industrial Revolution. The ever present but always weak descendants of the Industrial Revolution's first foes—Rousseau, the Romantics, Thoreau, and Emerson—began to find ready audiences for their values.

Values, like governments, are supported and cherished only so long as they are perceived to protect people from the things they fear—hunger, chaos, pain, sorrow, natural disaster, and other people. Two world wars and Vietnam showed the most destructive side of our society, and to some, its values seemed incapable of exercising control. More important than the causes of the wars were their scope and their demonstration of the destructive force of modern technology. For the first time, winning seemed to require destroying not only armies, but also entire cities, countries, and populations. In doing so, landscapes were also destroyed.

The story of World War I had its own frightening character. Unlike traditional narratives, it was difficult to see a clear beginning, middle, and end to World War I. American intervention seemed almost a protest against the formlessness, the senselessness of it. The destruction seemed ready to go on and on.

The Second World War reaffirmed the power of destruction demonstrated in the First World War, and refined and multiplied the technology. Unlike the earlier war, this one became a true world war and encompassed the globe. It ended, of course, with the weapon that epitomized all our fears about destruction—the atom bomb. Its power came from splitting the very building blocks of the universe. Although the bombs were small, the kind of power they unleashed, together with the effects of radiation, convinced everyone that humanity's tribes could now destroy not only each other but the entire planet. Whole civilizations could be "bombed back into the Stone Age." The world's imagination was prisoner to the prospect of bombs that could burn life from the face of the earth and even crack the planet in half.

Superimposed on the dark forces of the Industrial Revolution were the dark forces of the human soul symbolized by Hitler's campaign of extermination against Jews and other non-Aryans. Genocide was nothing new, but technology's ability to realize it on a large scale was. Equally powerful was the Nazi demonstration of the dark side of science with its bizarre experiments on human subjects.

In Vietnam, the high technology of America pitted against little people in black pajamas and sandals wheeling their war material on bicycles seemed to many young people to be a perversion of America's basic sympathy for underdogs and people who work the land. The Vietnam War became the first war in which environmental destruction was a symbol for the failure of industrial societies. After destroying, at least temporarily, vast areas of land, we lost to socialists employing an army of peasants. Waves of disillusioned intellectuals and middle-class young people launched a frontal attack on Western values. Some baby boomers became bombers.

The greatest horrors that Western technology could create had been demonstrated by war, but in 1962, the world learned that even plowshares were no longer safe. In *The Silent Spring*, a shy biologist named Rachel Carson warned that the abuse of agricultural chemicals was leading to the day when spring would come and no bird would sing. To paraphrase a famous line reportedly spoken by a U.S. officer about a Vietnamese settlement, we seemed bent on destroying the global village in order to save it.

A generation of baby boomers knew little history and less science. They lived in comfort that would have been the envy of Renaissance kings and queens. By contrast, current events and recent history seemed like the end of the world. While their parents blamed Hitler and communism, boomers blamed their parents and their culture. Despite the popularity of the word *counterculture,* the rebellions that started with Elvis and rock and roll in the late fifties and climaxed with the anti–Vietnam War movement in the seventies had no alternative culture to offer. If they offered anything, it was a mishmash of old ideas borrowed from shallow reading of oriental and primitive religions and an almost complete ignorance of past and present life in the cultures dominated by these religions.

The multicultural movement that began in the late 1980s can be justified as an attempt to widen the resources of Western culture, but it is really just another expression of disappointment and hostility. It picks up the environmental movement's idealization of primitive cultures and oriental mysticism and expands it to anything and everything non-Western.

A NEW BEGINNING

Every society steers its course through history guided by four main intellectual instruments: its map of history, its understanding of man's place in nature, its definition of itself, and its moral values. So long as the instruments that help a society survive and feel comfortable in the world are functional, the fundamentals of culture do not change. More simply put, we live by our myths. For an untouched tribe in the remote Amazon basin, a world run by terrifying humanlike gods and godlike animals provides all the story they need to survive in their specialized place and fulfill their limited needs and desires. In the civilized world, science has pushed us far ahead of our old myths.

The age of discovery that began with the Renaissance has been widening everyone's world for almost seven hundred years. It reached maturity in the global village, with modern communications and transportation and with the recognition that the most important environmental problems are worldwide. Like the Japanese finger trap, the more nations try to extricate themselves from the net of the world's economic, environmental, and political forces, the tighter these forces draw around them, or at least the more nations recognize the consequences of withdrawal. Only desperately poor and isolated countries stay aloof—Albania and Burma, for instance.

The myths that must sustain an industrial society in a global village beset by serious psychological and environmental problems must be more useful and complex than those of a rain forest tribe, no matter how much that society longs for primitive simplicity and stability. We must absorb all the revelations of science into their story of who we are. As we approach the twenty-first century, it seems clear that the myths of the Renaissance and the Industrial Revolution are no longer adequate guides for a global society.

The most demanding revelation of our time has come only in the last half of this century. If the Renaissance gave us the sense of true dominion, only the events since World War II have revealed that our dominion has become the power of life and death over nature as we know it. This is a big enough challenge, but it comes along with a crowd of others that demand we change our vision of the world.

Two sets of events are shaping our new renaissance. The first set is familiar to anyone who knows how classical civilization emerged or how the medieval era ended. The events are part of the historical pattern that marks this kind of huge cultural change. The second set of events are those unleashed by new freedom of inquiry and discovery. They are the events that reshape our way of seeing the world and begin to build a solid matrix of new values and new institutions.

SIGNS OF CHANGE

THE BREAKUP OF FEUDAL STATES

The modern Renaissance, like the Renaissance of the fourteenth century, began with the breakup of feudal states. The collapse of the Soviet economy, the mass exodus from East Germany, the dismantling of the single-party state in Poland, and the democracy riots in China were the first rubble from feudal states breaking up. The environmental importance of this phenomenon is its potential for unleashing the power of human creativity and thus our ability to adapt. If we do not squander the "peace dividend" paid by the end of the Cold War, we will be as rich as Isabella's Spain, Medici Italy, or Elizabethan England.

GREATER POWERS OF HUMAN OBSERVATION

When the second wave of discovery was in its infancy, the European Renaissance dissolved centuries of superstition and dogmatic thinking. It encouraged humankind to look at creation without preconceived ideas. The new tools were spectacular—the microscope and telescope, experiment and calculation. Its thought was daring: human beings had been created with unalienable natural rights to pursue knowledge and happiness. These ideas and the tools available broke old bonds by daring humankind to look directly at the bits and pieces of the universe from cell to galaxy. Our new knowledge filled us with self-importance. "What a piece of work is man," Shakespeare declared. This new self-confidence made man "the measure of all things" and set off an explosion of discovery.

An IBM executive told me a few years ago that fiber optics and computers could allow us to transmit huge amounts of data in milliseconds but no one yet had a use for such speed. Those uses have now begun to unfold and will continue to unfold for years to come. And fiber optics is only one of many new tools we have for examining and conveying information about the universe. These tools, as surely as the microscope and the telescope, will radically change our ideas about nature and about ourselves.

An Orgy of Discovery and Settlement

As modern technology turned a historic corner after World War II, humanity's third wave of discoverers had begun to visit regions almost ignored since human beings first came to the Americas some 50,000 years ago. As in every age, discovery is led by adventurers, driven by curiosity and profit, and followed by tourists and new settlers. Like the previous two waves, this one carries both familiar and new dangers. Yet it is unique in its ability and will to understand what it is about to exploit. Not quite knowing what would be most important about this era, word makers coined many names—the High Tech Age, the Rocket Age, the Space Age, and the Electronic Age. It is becoming ever clearer, however, that above all else, we are living in the Information Age. From satellites to deep-sea submersibles, we can probe on both the grand scale and the atomic scale as never before. In a matter of months or a few years, we can record more observations and make more measurements of a single region than were made of the whole world by every scientist who ever lived before 1950. More importantly, we can command our computers to organize and analyze our observations and measurements.

Our technology has armed us with unprecedented power. We have the ability to settle in comfort on the moon or at the bottom of the sea. We have the power to extract a few ounces of gold dust from tons of ore, to round up schools of fish by radar, and to turn tides into electricity. Perhaps most important, we have a new kind of knowledge that gives us not only the hindsight of history, but also a foresight that lets us see the consequences of our actions.

New Applications of Human Powers

The Industrial Revolution magnified human powers on a scale unimaginable at the beginning of the Renaissance. In the main, however, it used techniques and resources that were known—wheels, gears, steam, water, organic fuels, and wood and masonry construction. The Industrial Revolution was the climax of Renaissance science, politics, and economics. Almost all of its technology focused on magnifying human power, producing more goods with less labor and capital, and generally increasing the world's wealth. In the last half of our century, as affluence has allowed us to question the applications of technology, science is offering a new and very different set of powers.

The list of new technologies runs from gene splicing to synthetic fabrics that make Superman seem weak. How new are these powers? Simply ask yourself which of today's important technologies could have been imagined by the average educated citizen fifty years ago. Genetic engineering, personal computers, automatic language translators, earth-penetrating radar, solar-powered homes, electromagnetic resonance body scanners, fiber optics, laser surgical and surveying instruments, holograms, camcorders, magnetic levitation trains, synthetic insulin, oil-eating microbes?

The most dazzling power is access to the information generated by the new technology. As I wrote this book, I sat in my cabin in the forest in North Carolina, and over two small wires, my computer traveled through data bases around the world—libraries in Australia and Hong Kong, the National Science Foundation archives, documents from Russia's once closed city of Gorky (now Nizhny Novgorod) and its prison camp capital, Magadan. Reference volumes that once cost thousands of dollars if they were in print now exist on a single compact disk. Fiber optics will soon make all the world's literature, cinema, video, art, and music available by computer-telephone. The computer and satellite have created a global democracy of information and culture. Our Renaissance will use these and other new tools to change the way we live and think as surely as the first Renaissance did.

A WIDER INTERNATIONALISM

Nation-states and nationalism emerged from the Renaissance with greater size and durability than ever before. Equally important were the international dependencies (and enmities) forged as nations reached farther abroad for more varied and cheaper resources. In the 1990s, another leap in product complexity as well as environmental challenges has drawn nations together. Computers assembled where I live in North Carolina have components from seventeen countries. Athletic shoes sold in Russia are made from tops created in Indonesia and soles from Mexico. Pollution is also a multinational phenomenon. If global warming proves to be a fact, its contributors will come from every country in the world. The Renaissance voyages gave a great boost to international laws for sea transport. We now have international laws on the exploitation of ocean resources and an

international whaling commission. Agricultural and pharmaceutical companies have begun to forge agreements with tropical countries on the use of
genetic materials. For better or worse, we have global agreements on biodiversity, the production of ozone-threatening chemicals, and greenhouse
gases.

CHANGING THE HUMAN MIND

A NEW WAY OF PUTTING THE WORLD TOGETHER

We continue to examine the universe in ever finer detail, but an entirely
different kind of activity marks the true revolution of our time. Our age is
putting the pieces back together, allowing us to see ourselves and our
world as a whole. We have come to see life as a web, a woven fabric. Suddenly, so many pieces that seemed only beautiful or curious have become
meaningful. Just as Copernicus and Galileo removed Earth from the center of the universe, modern physics has shrunk us even further in relation
to the universe and our ability to understand it.

On the Sistine Chapel ceiling, Michelangelo painted the gesture that
became a symbol of the Renaissance—God reaching out his hand as if to
transfer his power to the awakening Adam. In our time, the symbolic
image has become the vision of a small blue and white planet in the
immense blackness of space. Humanity has not had a vision as singular and
persuasive, and perhaps as unifying, since times when political power and
religion were one and doubt was punished by exile or death. We know the
consequences of our worst mistakes. — do we?

REDEFINING HUMAN NATURE

While science has made us grow smaller in relation to the whole, we have
also grown closer to other forms of life. Biology and ecology have provided
an unexpected aura to our position among living beings. While we have discovered common bonds with sea slugs, squid, and fungi, we have also recognized the unique power of the human brain.

Both micro and macro biological explorations have led to a profound
rethinking of human nature. At the macro level, the behavioral links
between humans and animals explored by paleontologists and sociobiolo-
gists have destroyed forever the comfortable sense of special creation. For
some, this new kinship is like discovering that Rasputin, Hitler, and Charles
Manson are cousins. Just when we had grown accustomed to, and even fond

of, our apelike and monkeylike ancestors, we find out we are related to sea slugs. For others, it explains the link they always knew existed between themselves and, say, their dog, cat, or horse. It has refueled the animal rights movement. The closer the relative, the better protected it is. Anything that has eyes and can look back at us and blink is sure of protectors. The more human it looks, the better. Even E.T. would not have sold a single video cassette if he had stood on four legs and been eyeless.

Mitch Sogin at the Marine Biological Laboratory in Woods Hole, Massachusetts, heads the Center for Molecular Evolution, where he has used sophisticated genetic analysis to travel back in time to visit what he believes are the organisms in which most of our own biochemistry and physiology were pioneered. This search for the biological "big bang" began with Aristotle and continues in what appears to students as biology's most boring discipline—taxonomy, or systematics, the classification of life into orders, families, genera, and species.

Taxonomy is actually a form of biological history. Aristotle decided the world contained four hundred to five hundred forms of life. Until modern technology gave us electron microscopes, computer-enhanced light microscopes, and ways of reading the genetic code, scientists had to trace the family tree of life using relatively crude observations of similarities and differences. The technology available to Mitch Sogin has made a radical change not only in our definition of plants and animals, but also in what we think our ancestors were and how we fought our way up from primordial slime to big city lights.

Sogin began his career in molecular biology, but soon discovered "what a gold mine there was in eukaryotes,"[3] single-celled organisms with nuclei. Sogin's enthusiasm for this world known only to people with microscopes reminds me of L. L. Larison Cudmore's lament for the loneliness of the cell biologist:

> We are a sad lot, the cell biologists; like the furtive collectors of stolen art, we are forced to be lonely admirers of spectacular architecture, exquisite symmetry, drama of violence and death, noble self sacrifice and, yes, rococo sex. All found in the world of the cell. Cells have everything. But visibility.[4]

When Sogin began to find traces of our family tree in the biochemistry and architecture of eukaryotes, he asked himself, "Why would anyone want to work on anything but protists because they are so diverse?"[5] Among the brilliant and wonderfully formed protists, Sogin has discovered what he believes are the farthest reaches of our family tree. Constructing the tree now becomes a matter of filling in the trunks and branches between us and them. Redrawing the tree will redraw our sense of who we are and where we stand in nature's scheme.

The most profound discoveries about the nature of learning, vision, and hearing have been made not with monkeys, cats, or pigs, but with invertebrates, and not even terrestrial invertebrates. Eric Kandel and Daniel Alkon, working with flubbery, slimy sea snails, have traced neural and chemical mechanisms of learning that appear to be valid for human beings. Robert Barlow at Syracuse University, working with horseshoe crabs, the "30-million-year-old fossil," has outlined how the brain and the eye communicate. Many genetic discoveries with fruit flies and other relatively simple organisms have reshaped our thinking about human heredity.

Recently, a report in *Nature* said that it is not unusual for plants to carry the molecule that we used to think was uniquely inherited by animals to carry oxygen in blood. While scientists have known for some time that hemoglobin molecules can be found on the root nodules of some plants, it was thought to be a rare quirk—something like the platypus, or a monkey at a typewriter banging out a quotation from Shakespeare. They now think that both plants and animals have inherited an ability to produce and use hemoglobin for respiration.[6]

The importance of discovering these new connections is in the newly visible bonds of kinship between humans and nature. This real world makes the spirit world of primitive people seem to be what it is—the best they could do with what they have, but childlike and exceptionally naive. It's a dazzling fantasy world far from nature's much more stunning reality.

A NEW KIND OF ECONOMICS

We talk now about the "human costs" of environmental problems. We have begun to expand our bookkeeping to include not only new costs, but also past costs that were ignored or paid for in the lives of people considered unimportant. As international jingoism subsides, the lives of other human beings become more important. We may lament the spread of a common culture across the world's rich pattern of traditions, but the commonality may bring us closer together. The sudden acceleration in international economic ties that lies behind the homogenization of culture also makes war less likely and cooperation on environmental improvement easier. The North American Free Trade Agreement has already provided political and economic incentives for stronger environmental programs in Mexico.

Together with human costs, we have begun to measure natural costs. Conservatives and liberals alike agree that our current markets do not reflect the full costs of our goods and services. Here conservatives have taken the lead in proposing programs to remove subsidies and have each product and service pay its own way. The earliest and easiest of these measures were the "user fees" pushed by the Reagan administration in the early 1980s. Although liberals generally oppose privatizing resources, they have slowly begun to endorse privatizing even the air by selling "pollution rights." They might hesitate to call it privatization, but if we set clear limits on the pollution of water or air in a certain region and allow the rights to pollute to be traded, we are selling private rights to what were once common resources. If we insist that the market recognize environmental costs, we will revolutionize economics.

NATURE ALMOST FOR ITS OWN SAKE

Our new sense of the complex interactions of nature has provided a new and unique reason for preserving natural areas. In the political arena, the reasons given for saving wilderness, rain forests, and other exotic ecosystems are often economic. These regions store disease-resistant genetic materials that we might use to improve domestic crops. They could provide cancer-fighting drugs. We can develop new products from their resources— buttons from tagua nuts, unusual fruits, new lubricants, and even fuels from the sap of trees. Few economists take these proposals seriously and few scientists believe they are the best reason for saving unique natural areas. The real reason for doing so is that high-tech science is revealing that we have vastly more in common with the rest of life than we ever thought.

Saving nature for this reason also exhibits our self-interest, but it is quite different from the self-interest of people who believe that rocks, trees, and animals are gods and that personal pain and suffering will follow any sins against nature. That kind of kinship is an adult version of what I have called "teddy bear nature," or humanized nature. Our new appreciation of nature also differs greatly from the much more economic and materialistic interests of primitive people living hand to mouth.

The motivations behind the Endangered Species Act are multiple. In great measure, the act serves two strategic interests of the environmental movement. First, it is a Trojan Horse wheeled into the legal system to take vengeance on industry and development. (Why else is a law with no results so popular?) Second, it is a strategy for appropriating wild areas for environmentalists rather than other users. This is a strategy that has nothing to do with saving nature, but everything to do with competing economic choices. Environmentalists want someone else to pay for saving natural areas they value for personal reasons. Industry, ranchers, loggers, and farmers want their share no matter what environmentalists lose. However, the Endangered Species Act also writes into law our growing recognition that when we lose a species, we often lose part of our own story, our own biological genealogy and family album.

THE DISAPPEARANCE OF GOOD AND EVIL IN NATURE

The power of technology to extend our perceptions of the natural world has challenged even our strongest moral principles. "Thou shall not kill" is still a sound idea, but because we can see into wombs, fertilize human eggs in a test tube, and pump air and blood into people after their brain has died, we are now arguing over the very definition of life and killing. We know we should not put poison in each other's food, but refinements in chemical analysis of living tissue show that our bodies are full of poisons in small amounts all the time, and that even the fruits and vegetables we eat expose us to their own natural toxins. A chemical that helps in small quantities kills in large ones. Gone are the days when we could detect only fairly large doses, and a law such as the Delaney clause could ask for banning any chemical that causes cancer in animals. We can now detect chemicals that exist as only a few molecules among a trillion others, the equivalent of a teaspoon of toxics distributed in the waters of the Gulf of Mexico. Our cells and their reproductive and hormone-producing mechanisms are bathed in a sea of chemicals, many of them toxic and carcinogenic. Many change cell functions even if present only in parts per trillion. Instead of simply accepting or banning chemicals, we are using sophisticated biological investigation and computer calculations to measure risk. We are going to have to decide how much risk is too much, and even how many deaths we will tolerate.

We have realized that good and evil exist in our minds and our moral lives, but not in nature. With the exception of a relatively few disease organisms, there are not many microbes, viruses, plants, animals, or chemicals that hold up before us their smoking gun of biological mayhem. We have become pretty good at policing the smoking guns. Medically and philosophically we have only just begun to deal with those organisms and chemicals that are sometimes good, sometimes bad, whose dangers are occasional or present only in a crowd where they cannot be singled out.

ORDER IS HUMAN, BUT NOT NATURAL

Nature once seemed to be a well-ordered system. If it was not as simple as the divinely designed clock of the eighteenth-century physicists, then at least it had a recognizable balance we could describe and maintain. But no. Our investigations of the gas trapped in ice cores, radioactivity in cave accretions, pollen in lake sediments, and other clues tell us that nature is neither balanced nor cyclical. It is like a highway through time, and along with every other creature on the earth, we are headed for unpredictable surprises, sudden forks and intersections, and a destination entirely unknown. Instead of chaos being what we don't know, chaos is what we do know. While we depend on order, nature is disorderly. Therein we must accept a basic disagreement. Our challenge is how to disagree with nature in a civilized manner.

CAUSE AND EFFECT ARE OBSOLETE

As the complexity of nature becomes clearer, we find less and less use for the old standby of both morality and science—cause and effect. We have seen two quick and simplistic reactions as both liberals and conservatives run away from the challenges of this realization.

Liberals began to act as if all values are relative. No one can judge right or wrong, or place the merits of one culture above another. Environmentalists decided that they can simply choose the cause and effect that serve their cause. If sulfur emissions from coal-burning plants cause some acid precipitation, they become the cause of acidic lakes. If capitalist countries consume tropical hardwoods and Latin American hamburger beef, they become the cause of deforestation. If dioxin alters a cell's reproductive functions and slightly depresses the immune system, it becomes the cause of cancer.

The predominant conservative reaction to the loss of cause and effect has been a worldwide return to religious fundamentalism and puritanism that mirrors environmentalist puritanism. Is it a reaction to the amoral world of liberal society or a fortification against the challenges of science?

Some conservatives refuse to allow the light of new knowledge to shine on fixed judgments. A clear example is the failure to admit the possibility of a biological basis for homosexual behavior. Even though every society chooses its permissible behaviors, if some homosexual preferences are biologically programmed, is it logical to insist that homosexuality is a matter of moral choice? Nor have the most militant anti-abortion, pro-life activists dealt with the problem of when life begins. If it begins at conception, then why not before conception in the creation of egg and sperm cells? If it begins when a soul enters the product of conception, when is that and how do we know? At the other end of life, must we now spend millions of dollars keeping alive people whose brains no longer think, whose bodies do not function without help? And if we are not obliged to help sustain life to the limits of our technology, then is it also okay to deny support to infants who need it? Even more egregious is the conservative approach to toxins. Since the multiple interactions of chemicals in our bodies blur clear cause and effect, many conservatives say that a chemical such as dioxin is risk-free unless we spread it on school lunches. They call talk of *probable* effects "technobabble."

Until these evasive reactions of both liberals and conservatives give up the stage, the passing of cause and effect will not have its full impact. When we do absorb the full meaning, we will find ourselves in a new world. It will have less certainty and more freedom. We will realize that living comfortably with nature, whether it is the nature of our internal chemistry or the earth's atmosphere, will require our active participation in managing the planet's life-support systems. It will not be a simple matter of saying what we can and cannot do, a simple division of the world into good and bad chemicals, energy forms, development styles, and agricultural practices.

Yes and no commands should be reserved for the most obvious moral principles. "Thou shalt not kill" is still good advice, even if we have separated the judgment of the crime from our judgment of the criminal. Applied to the management of economies and nature, such authority yields the kind of centralized micromanagement that kept totalitarian countries in political and economic slavery while other countries enjoyed unprecedented growth.

NATURE HAS NO WILL BUT OURS

If the vast, apparently lifeless universe is nature, then life is hardly one of nature's preferences. Life is but a millisecond in universal time, hardly nature's favorite pastime. So what does nature want for itself? If it wants anything, we don't know. Chaotic nature shows no clear development plan, unless it is a general trend from simplicity to complexity. Even that may be an accident. So how should we treat nature?

The environmental movement has always maintained a self-serving ambiguity about whether its concern is the survival of nature or humanity.

Nature, of course, will survive in one of its many forms or in a new form despite the worst we can do. If environmentalists care about nature, they care primarily about the present state of nature, the nature they grew up in and like, or perhaps an idealized version that exists in wilderness or in some less populated time. But to care for just this state of nature is to care for a condition that is very unusual and short-lived in the history of the universe or even the much shorter history of life on earth. In other words, it is not really caring for nature, but for our own preferences in nature.

The debate in our society is not really between those who care about nature and those who do not. Everyone has a preferred environment. Less than half of 1 percent of the world really wants to live in a wilderness, or even in a cabin at Walden Pond. Put aside all the elegant essays and wilderness treks and it is clear that we all prefer nature to be subservient to our own interests. The debate is over how to manage nature for human purposes.

The more serious part of the debate is which management philosophy will ensure that nature remains subservient to us, now that we have achieved real dominion. How do we keep the oceans supplying us with fish? How do we keep the atmosphere from reacting too extremely to our use of oil, gas, and aerosols? If we want to cut back on the carbon dioxide in the atmosphere, where is the best place to store it—in the tropical forests, in the soils, in agricultural crops, in manufactured wood products? We know how to preserve grizzly bears, pandas, elephants, and rhinos, but we have to ask if we really want to, and if the answer is yes, how many we want and where we want them. Since we have surrounded their habitats with our own civilization, there is no returning the decision to nature. It is up to us to decide whether they starve to death, get killed by poachers, or are managed by the most scientific and humane methods we have.

Rutgers philosophy professor David Ehrenfeld, who specializes in environmental ethics, asks of his fellow environmentalists a fundamental question for which neither liberals nor conservatives have yet found a guiding principle.

> We do not know how many species [of plants] are needed to keep the planet green, but it seems unlikely to be anywhere near the more than quarter of a million we have now. Even a mighty dominant like the American chestnut, extending over half a continent, all but disappeared without bringing the eastern deciduous forest down with it. And if we turn to the invertebrates, the source of nearly all biological diversity, what biologist is willing to find a value—conventional or ecological—

for all 600,000 species of beetles? . . . In short, the ones most likely to become extinct are obviously the ones least likely to be missed by the biosphere. Many of these species were never common or ecologically influential; by no stretch of the imagination can we make them vital cogs in the ecological machine.[7]

Natural Resources Are Almost All a State of Mind

Oddly enough, environmentalists believe that nature can manage animal populations, but they insist that human populations be subjected to limits set by government officials—consulting with environmental experts, of course. A world of 60 billion people would almost surely be unpleasant by today's standards, but not unsustainable in the next century. Methods of housing, feeding, and transporting people will undergo a revolution as unforeseen as fiber optics. Fortunately, population growth is likely to level off between 12 and 15 billion midway through the next century. For a little perspective, we should note that all these people could fit in an area twice the size of Rhode Island. That would be shoulder to shoulder, but the rest of the world would be empty. More rationally, we might consider that they could live in a futuristic megacity or "arcology" the size of Texas, with the rest of the world still empty. I would not want to live there, but I offer the picture to point out that even 15 billion people would not populate the world as densely as ants in an anthill.

Besides, we still have large areas of land unpopulated. Cities in the deserts, tundra, mountains, and taigas would not disturb anything more valuable than we are already disturbing. Technology allows us to live in reasonable comfort and harmony with hostile environments. Before the invention of fire and animal skin clothing, humanity did not spread out into the subarctic. Possibly not even out of the warmer temperate zones. Only fifty years ago, modern urban concentrations were not viable in the American South or equatorial regions because there was no affordable air conditioning. The humble cornerstone of the American Sunbelt's office and industry boom is the air conditioner, its anthem the hum of condensers. Today we take all these habitats for granted. Tomorrow it may be arctic suburbs or floating ocean cities or lunar colonies.

As the world's population growth slows, economic growth can continue to supply more goods and services per person. When it gets too expensive to supply more copper or more oil, science will provide the same benefits from another resource. Those who say this is not possible defy both history and human nature. In the past, economic growth was limited by the resource base. Substitution was as slow as population growth. If moving to new lands was cheaper or easier than inventing new products, people moved. Polynesians did not sail the vast and dangerous Pacific just to sing sea chanties. Asians did not cross the Bering land bridge and walk south between the walls of glaciers to enjoy the view. Until the Renaissance and the Enlightenment combined science, economic power, and creative freedom, most societies did not have a systematic way of solving resource shortages except through war or migration. The deliberate exercise of imagination to find new resources was sporadic, stimulated mainly by the challenges of being pushed into an unfamiliar or inhospitable environment.

Long before Europeans arrived, inhabitants of North America began fighting with each other because their technology limited their access to resources. Their values no more prohibited genocide than the values of European Christians prevented the Crusades. The Yamana by the Strait of Magellan did not take up residence in the gales, freezing rains, and snows because they had come to the end of a happy pilgrimage toward Antarctica. They had been forced there by stronger tribes.

When the Europeans met natives in the Americas, the natives had come to the end of their ability to replace exhausted resources by migrating. Almost constant warfare among tribes had begun before the Renaissance in Europe propelled conquerors toward the Americas. The end of one era in human adaptation had met the beginning of another.

At the end of the Middle Ages and the beginning of the Renaissance, around 1400, the world population began the steep rise that has climaxed in the present "population explosion." During this time, though, the growing population did not exhaust resources; instead, humankind expanded the resources available for everyone. The one resource most essential to productivity, comfort, and leisure is a good case in point. I mean energy. For thousands of years, almost all energy for cooking, smelting, making glass, and producing steam came from forests. Wood is a renewable resource if well managed, but it never was. In addition, burning wood with its carcinogenic hydrocarbons and lung-clogging particles meant multiple health problems for people living in crowded tents, huts, cabins, and caves. For ordinary work and transportation, human muscle was used most for thousands of years until we domesticated animals. These animals, of course, required huge areas of forest to be cleared and cropland to be diverted from feeding people. In the eighteenth century, we expanded to coal. About 1900, oil came in, then gas. Then nuclear fission. Despite real and potential problems from nuclear accidents and atmospheric pollution, most people in the Western world breathe cleaner air than people did in the days of wood and coal. Today we stand at the beginning of efficient ways to harvest solar, wind, geothermal, and biomass energy. We use less and less energy per dollar of goods and services.

To realize how different our economy is and the importance of our new technological tools, we might consider what we are spending money on. Just after World War II, about 40 percent of the cost of consumption went to buy raw materials such as copper, rubber, coal, and steel. The business analyst Peter Drucker estimates that today only 3 percent of the cost of consumption goes for raw materials. The rest goes for interest, research, software, design, and other economic and intellectual products.[8]

Warnings of danger always appear in exceptions to trends, and they should be carefully observed, but the results of civilization's development are measured in the trends themselves. Despite numerous setbacks, all the trends demonstrate that the forms of human creativity nurtured in Western science and economics have been generally good for humanity. They have been especially good for those nations whose political systems embraced them. In these countries, by almost anything that can be clearly measured, life has improved tremendously for citizens in all parts of society. Two clear examples:

- *Life expectancy* in Europe and America has risen from less than twenty years four thousand years ago to thirty-three years in 1700 to about seventy years today.
- *Leisure, play, and creative time have increased.* Primitive people spent a third of their life's hours working, compared to 10 percent for citizens of modern industrialized society. A sixth of life's hours in primitive societies were free for play, leisure, and creative pursuits, compared to almost 40 percent in the industrialized world. In developing countries, a tenth of a person's life is consumed by serious illness.[9]

The Challenge Is Abundance, Not Scarcity

For most people who lived through the Great Depression of the 1930s or who were raised by parents who lived through it, those years reinforced the world's oldest threat—scarcity and poverty. They were the legacy of thousands of years of history during which most people lived on the edge of disaster and hunger. For hundreds of years, we continued to use the same basic pool of resources, and as always, logic and experience dictated that their end would come—and ours soon after.

The idea that we live in a static world with clear limits to progress and to the ways we can adapt dies hard. Like Rasputin in his last hour, the idea is full of poison but still wants to call the tune and dance. The fears of Malthus put on the power of the computer in 1972, when the Club of Rome predicted that industry would soon begin to exhaust the world's copper supply, and civilization, stripped of its communications lines, would collapse. But then came fiber optics and cellular telephones, and microwaves bounding off of orbiting communications satellites.

Shortly after World War II, an eccentric navy veteran of World War I saw the essence of how his century could spread unprecedented abundance throughout the world using nothing but the resources already at hand. Buckminster Fuller, an industrial designer, writer, futurist, and inventor of the geodesic dome, recognized that the material progress of the industrialized world was its ability to provide more and more human advantages from fewer and fewer resources. His domes, for instance, covered several times the space of a traditional building with a fraction of the usual materials. Similarly, a quarter-ton communications satellite can do the work of 150,000 tons of submarine cables.

Fuller recognized that the ability to extend the benefits of the world's resources to more and more people had begun to accelerate rapidly, and it contained the potential for improvements in life that would have been the envy of the early utopians. He estimated that from the time of his birth in the late nineteenth century to 1965, industrialization had brought to 40 percent of the world's people "a personal standard of living and health superior to that ever enjoyed by a pre-20th century monarch."[10] It is easy to argue with the exact numbers, but the point is clear and valid. Early attempts at utopia, Fuller said, had not only come too early, but also "all of the would-be Utopians disdained all the early manifestations of industrialization as 'unnatural, stereotyped, and obnoxiously sterile.'"[11] Many environmentalists still feel this way. While they lionized Fuller briefly in the mid-1970s, they soon discarded his optimism for the pessimism of Thoreau. From the White House to Greenpeace, environmentalists have not yet learned that we do not adapt and survive by imitating plants and animals, but by becoming more human.

As in any Renaissance, groups often at odds with each other are ultimately contributing to a consistent process of change. So long as we have a debate between those who would turn back and those who would go forward, our Renaissance is still alive. To carry on a fruitful discussion, we will need people who are self-confident enough to appreciate worthy opponents; open-minded enough to welcome diversity; and critical enough to detect distortions of fact, reject sloppy thinking, and resist glittering illusions.

In 1966, the environmental movement was presented with two quite opposite visions of the future. Buckminster Fuller wrote, "For the first time in history it is to be assumed now and henceforth that it is normal for man to be a physical and economic success."[12] For Fuller, the marriage of science and technology in the Industrial Revolution had given us the means for success. But for an equally influential historian, this marriage put the sword of destruction in the hands of the world's most environmentally immoral religion. The day after Christmas, Lynn White delivered his famous environmental condemnation of Western society to the American Association for the Advancement of Science. He laid the blame squarely on Christianity: "Christianity, in absolute contrast to ancient paganism and Asia's religions . . . not only established a dualism of man and nature but insisted that it is God's will that man exploit nature for his proper ends." In case the point wasn't clear, he added, "By destroying pagan animism, Christianity made it possible to exploit nature in a mood of indifference to the feelings of natural objects."[13] Our ecological crisis would get worse, he predicted, until we stopped putting human needs ahead of nature. Instead, we must "substitute the idea of the equality of all creatures, including man, for the idea of man's limitless rule of creation."[14] White's speech, quoted and paraphrased thousands of times by environmentalists who do not even know his name, has shaped the fundamental theology of the environmental movement.

White's argument that Christianity's values are more destructive than any other culture's does not stand even a quick reality check. But he also lays down a truth that can help us understand where we are going. "What people do about their ecology," he said, "depends on what they think about themselves in relation to things around them."[15] Therein lies the key to our future. It is clear that if we condemn ourselves as White and other environmentalists do, we will abandon our most precious gifts. We can only preserve nature as we prefer it, and achieve the kind of peace and comfort Fuller describes, if we embrace our powers.

We do have almost "limitless rule of creation." We should attempt to change creation for our own convenience. Even White begins his famous speech by saying, "All forms of life modify their contexts." And he could have added that in a world filled with life, the advantage of one species is always the disadvantage of another.

Many of us will always regret the passing of the wilderness and the frontier, the days when humans tested personal courage and wisdom individually against the forces of nature. People will still climb sheer cliffs and try to walk across Antarctica, but such tests have become increasingly voluntary and recreational. We have entered an era when nature no longer tests us. From now until the end of civilization—or our species—we will test nature. Dominion is ours. As we test nature, we will also test ourselves and the very limits of human wisdom.

We have unleashed dangerous forces in the world. We created them and we understand them a lot better than primitive people understood the weather or the waxing and waning of the animal populations they hunted. We are in no worse position than humans ever were. We have lost some things and gained much. We may regret the loss of Glen Canyon or the forests of Manhattan Island or the clouds of passenger pigeons that once darkened midwestern skies. But we don't regret the passing of yellow fever, malaria, or bubonic plague. Few people really want a simpler life. We may turn down the thermostat five degrees, but we won't take out the heat pump. We all want more, for ourselves and for the rest of the world, because that is the way to peace. And we know it is possible. Now that we know the freedom of the human mind is more important than the quantities of any natural resource, no insurmountable obstacle exists to continued improvement in the quality of our lives and the way in which we manage the natural world.

NOTES

The works cited appear either in the bibliography of important books or the bibliography of works cited and used.

CHAPTER 1: CONFESSIONS OF AN ENVIRONMENTALIST

1. Carl Pope, Sierra Club direct mail communication, November 1993.
2. Stutz, "Landscape of Hunger," 54–63.
3. See Bibliography for Bramwell, Simon and Kahn, and Botkin.
4. Gore, *Earth in the Balance,* 274.
5. Ibid., 177.
6. Stroup, "Global Warming."
7. Ames, "Misconceptions," 34.
8. Brimelow and Spencer, "You Can't Get There from Here," 60.
9. Ibid.
10. Patton, "Who's Endangered?" 1.
11. Mann, "Extinction," 736.
12. Hardin, *Nature,* 275.
13. Rosenberg and Birdzell, *Economic Transformation,* 3.

CHAPTER 2: THE SEARCH FOR AUTHORITY

1. Paul Johnson, *Intellectuals,* 339.
2. Krug, "Corrosion of Science," 30.
3. Ibid.
4. Brooks, "Conservation Revolution," 36.
5. C. Vann Woodward, quoted in Reiger, *American Sportsmen,* 13.
6. Snow, *Environmental Movement,* 76, 71.
7. Becker, *Heavenly City,* 102.
8. Panczenko, "Ridiculous Degreed Ass."
9. Jevons, *Coal Question,* 164.
10. From the founding statement of the Audubon Society.
11. "Man the Destroyer."

12. Carson, *Silent Spring*, 103.
13. Ehrlich, *Population Bomb*, xi.
14. Smith, "Environment Since Industrial Revolution," 11.
15. Smith, "Carnival of Dunces," 30.
16. Wald, "There Isn't Much Time," 22.
17. Barnet, "No Room," 32.
18. Maddox, *Doomsday Syndrome*, v.

CHAPTER 3: AN OPPOSITION MOVEMENT IS BORN

1. Quoted in Hudson, *Landscape Linkages and Biodiversity*, 71.
2. Gore, *Earth in the Balance*, 230.
3. Marx, *Communist Manifesto*, 124.
4. Wollstonecraft, *Rights of Woman*, quoted in Noyes, *English Romantic Poetry*, 190.
5. George, *Life and Death of Haydon*, 67.
6. Malthus, "Essay on the Principle of Population," 51.
7. Ibid., 59.
8. Ibid., 38, 39.

CHAPTER 4: THE ROOTS OF ENVIRONMENTAL THINKING IN AMERICA

1. Quoted in Marx, *Machine in the Garden*, 41.
2. Ibid., 43.
3. Ibid., 42.
4. Ibid., 43.
5. Catlin, *Letters and Notes*, 261–62.
6. Quoted in Huth, *Nature and the Americans*, 317.
7. Reiger, *American Sportsmen*, 35–36.
8. Quoted in Holloway, *Heavens on Earth*, 128.
9. Thoreau, *Walden*, 383.
10. Ibid., 128.
11. Ibid., 424.
12. Ibid., 433.
13. Ibid., 127.
14. Ibid., 106.
15. Ibid., 120.
16. Ibid., 152.
17. Ibid., 109.
18. Ibid., 182.
19. Noyes, *English Romantic Poetry*, 373.
20. Holloway, *Heavens on Earth*, 80.

21. Quoted in Echard, *Organization Trends,* 14.
22. Wild, *Pioneer Conservationists,* 117.
23. Bailey, "Apocalypse Abusers," 3.
24. Daniel, "Revolution of the Right," 4.
25. *Blue-Print for Survival,* 5.
26. Lappe and Collins, *Food First,* 114.
27. Anthony, "Eco-Justice," 19.
28. Quoted in Foster, "Captain Planet," 8.

CHAPTER 5: SEARCHING FOR A NEW
SENSE OF THE SACRED

1. Marx and Engels, *Communist Manifesto.*
2. Bramwell, *Ecology,* 243.
3. Gore, *Earth in the Balance,* 259.
4. Witt and Steiner, *The Way,* 29.
5. Deloria, *God Is Red,* 365–66.
6. Johnson, "Machiguenga Manage Resources," 221.
7. Ibid.
8. Snyder, "The Etiquette of Freedom," 114.
9. Olcott, *Kazakhs,* 248.
10. Milton, "Civilization and Its Discontents," 40.
11. Molina, "Pilaga Seduction Songs," 43.
12. Pinkalla, "Pygmy Hippos."
13. Shele, 19.
14. Johnson, "Caring for the Earth," 22.
15. Ehrenfeld, "Desert Life," 8.
16. Ibid.
17. Ibid.
18. Redford, "Ecologically Noble Savage," 47.

CHAPTER 6: WINNING THE PUBLIC AWAY FROM SCIENCE

1. Sidey, "Opposing View," 21.
2. Brown et al., "World at Risk," 4.
3. Monastersky, "Deforestation Debate," 26–27.
4. Simmons and Kreuter, "Save an Elephant—Buy Ivory."
5. De Bell, *Environmental Handbook,* 163.
6. Snow, *Inside the Environmental Movement,* 7.
7. Mitchell and Stallings, *Ecotactics,* 151–52.
8. Ibid., 152.
9. Brooks, "Journalists and Others."
10. Krug, "Corrosion of Science," 29–30.

11. Krug, "Environmentalism," 44–46.
12. Ibid.
13. Ibid.
14. Ames, "Science and the Environment," 4–5.
15. Ibid.
16. Ibid.
17. Arnold, *Fear of Food,* 45.
18. Harte et al., *Toxics A to Z,* 283.
19. Arnold, *Fear of Food,* 97–98.
20. *Audubon,* 94–95.
21. Koshland, "Scare of the Week," 244.
22. Chakravarty, "Dean Buntrock," 98.
23. Babbitt, "World after Rio," 30.
24. Sitarz, *Agenda 21,* 33.
25. "Heidelberg Appeal."
26. Javna, *Fifty Simple Things.*
27. Allen, "School Children Learn Ecology," 1.
28. Javna, *Fifty Simple Things.*
29. Hocking, "Paper Versus Polystyrene," 504.
30. Dukes, "Evil Science," A27.
31. Weilbacher, "Earth Day," 35.

CHAPTER 7: NATURE AS WE WANT IT

1. Gore, *Earth in the Balance,* 11.
2. "Summit of the Arch," 1.
3. Botkin, personal interview, November 1992.
4. Nietzsche, *Beyond Good and Evil,* Section 9, 13.
5. Picard, "James Bay II," 15.
6. Botkin, personal interview, November 1992.
7. Thomas Berry, speech to North American Conference on Christianity and Ecology. Webster, Indiana, Aug. 14, 1987.
8. Stevens, "Balance of Nature?" 4.
9. Botkin, personal interview, November 1992.
10. Ibid.
11. Talbot, "Man's Role in Managing the Global Environment," 25.
12. Cronon, *Changes in the Land,* 11.
13. Gomez-Pompa and Kaus, "Wilderness Myth," 279.
14. Tilman and Wedin, "Oscillation."
15. Stevens, "Balance of Nature?"
16. Yoffe, "Silence of the Frogs," 64.
17. Ibid.
18. Botkin, personal interview, November 1992.
19. Ibid.

20. Ibid.
21. Ibid.
22. Center for Study of the Environment, introductory brochure, 1992.
23. Shoumatoff, *World Is Burning*, quoted in Baden, "Tropical Tragedy," 44.
24. Meadows, seminar at Amherst College, 1991.
25. Botkin, *Discordant Harmonies*, 192.

CHAPTER 8: WHO OWNS NATURE?

1. Ridley and Low, "Can Selfishness Save," 78.
2. Zwick and Benstock, *Water Wasteland*.
3. *Natural Resources Defense Council v. Callaway*, 392 F. supp. 685 (D.D.C. 1975).
4. Douglas, *Wilderness*, 146.
5. Brookes, "Glancing Geese," 112.
6. Stroup, "Endangered Species Act," 2.
7. Wright, "'Property Rights,'" 2.
8. Callahan quoted in Miniter, "Out to Change," 34.
9. Daniel, "Revolution of the Right," 4.
10. Reigle, "Environmentalists Concerned," 7.
11. Blessen, "Wetlands Policy," 25; and *Hoffman Homes, Inc. v. EPA*, 961 F. 2d 1310 (7th Cir. 1992).
12. Chafee quoted in Schneider, "Environmental Laws Face a Stiff Test."
13. Brookes, "War on Property Rights."
14. Echeverria quoted in Lehman, "Accord Ends," E1.
15. Meyerhoff quoted in Kaplan and Cohn, "Open Season," 70.
16. *Lucas v. South Carolina Coastal Commission*, U.S. Supreme Court, June 29, 1992.
17. Personal telephone interview, October 1993.
18. Sugameli quoted in Miniter, "Out to Change," 10.
19. Ibid., 9.
20. *Florida Rock Industries v. United States*, 21 Cl. Ct. 161 (quoting *Agins v. Tiburon*, 447 U.S. 255), 260–61 (1980).
21. Nelson, "Green Hills," 38.
22. Cuomo quoted in Drummey, "Land Grabbers," 25.
23. Reilly, quoted in "Review and Outlook."
24. Chase, "Property Rights," 4.
25. Stroup, "Voters Need," 4.
26. Ehrlich and Wilson, "Biodiversity Studies," 762.
27. Healey, *Competition for Land*.
28. Sandburg, *The People, Yes*, 72.
29. Borelli, "Whose Woods," 44.
30. Ibid.
31. Hair, "In Bear Country," 30.

32. Leal, "Unlocking the Logjam," G4.
33. Ibid.
34. Daly, "Free-Market Environmentalism," 175.
35. Quoted in Laffer, "Protecting Ecologically Valuable Wetlands," 11.
36. Corcoran, personal telephone interview, summer 1993.

CHAPTER 9: TECHNOLOGY TO THE RESCUE

1. Hoffman, "Ratepayers Beware," 3.
2. Ibid.
3. Browne, "Cooling with Sound," C7.
4. Locke, "Research Shows Consumers Concerned," 1.
5. "Beware of False Gods."

CHAPTER 10: THIS IS NOT THE END, BUT THE BEGINNING

1. Ehrlich and Ehrlich, *Healing the Planet,* 284.
2. Jackson quoted in Eisenberg, "Back to Eden," 89.
3. Personal conversation, Woods Hole, Mass., July 1993.
4. Cudmore, *The Center of Life,* 6.
5. Personal conversation, Woods Hole, Mass., July 1993.
6. "Blood from a Turnip," 26.
7. Ehrenfeld quoted in Wilson, *Biodiversity,* 215.
8. Drucker quoted in Libby and Clouser, "Population," 19.
9. Mahler, "People," 67.
10. Fuller, *Utopia or Oblivion,* 5.
11. Ibid., 289–90.
12. Ibid., 340.
13. White, "Historical Roots," 1205.
14. Ibid., 1207.
15. Ibid.

A BIBLIOGRAPHY OF
IMPORTANT BOOKS FOR
UNDERSTANDING
ENVIRONMENTAL THINKING

Bosselman, Fred, David Callies, and John Banta. *The Taking Issue*. Washington, D.C.: U.S. Government Printing Office, 1973. Funded by the Council on Environmental Quality, this book is a readable background on a topic vital to any effort to encourage environmentally sound land use. It covers the legal history of government taking of private property rights, but remains quite understandable for nonlawyers.

Botkin, Daniel. *Discordant Harmonies: A New Ecology for the Twenty-First Century*. New York: Oxford University Press, 1990. This often eloquent book has been generally ignored by the environmental movement because Botkin comes down on the side of hands-on management of nature. The book draws a sharp contrast between the human passion for order and nature's usually disorderly and unpredictable patterns. It is the first generally accessible book to assimilate chaos theory into environmental thinking.

Bramwell, Anna. *Ecology in the Twentieth Century: A History*. New Haven: Yale University Press, 1989. Bramwell's minutely documented history of green politics is the best of the scholarly histories of the movement. She demonstrates how a variety of movements fed on the myth of the "simple folk" and rural virtues, and how this fanciful revolt against industrial society aggravated some of the worst problems of this century, from Nazism to the glorification of socialism.

De Bell, Garrett, ed. *The Environmental Handbook: Prepared for the First National Environmental Teach-In*. New York: Ballantine Books, 1970. An anthology of essays prepared for the April 22, 1970, teach-ins that occurred around the country. Friends of the Earth sponsored this book, which sets out the environmental movement's core issues.

Epstein, Richard A. *Takings: Private Property and the Power of Eminent*

Domain. Cambridge, Mass.: Harvard University Press, 1985. This book became a powerful reference for Reagan-era attempts to soften the impact of environmental regulation on private property. Epstein's work continues to fuel legislation requiring government studies of regulatory costs to private property and of compensation for lost value.

Hardin, Garrett. "The Tragedy of the Commons." *Science,* December 13, 1986. This essay on the abuse of commonly shared resources develops one of the most widely discussed concepts linking ecology and economics. Environmentalists have struggled to absorb Hardin's recognition of the crucial role of self-interest into their more idealistic view of human nature.

Leopold, Aldo. *A Sand County Almanac* and *Sketches Here and There.* New York: Oxford University Press, 1987. Leopold has become the scientist counterpart to Thoreau for the environmental movement. A wildlife biologist and sportsman, Leopold's details are out of date scientifically, but his big picture of an intricately balanced nature still inspires environmentalists. Leopold's insights into what we should look for in order to understand nature and how to do so are still good guides. While his writing is not as polished as Thoreau's, it shines, and it is informed by the depth of scientific understanding instead of Thoreau's shallow oriental mysticism.

Lovelock, J. E. *Gaia: A New Look at Life on Earth.* New York: Oxford University Press, 1979. Often to its author's discomfort, this book has become the source of a new religious vision for many environmentalists. Lovelock, now an independent scientist in Britain, began developing the Gaia metaphor— earth as a single organism—while working on ideas for tests that would determine if life does or ever did exist on Mars. In scientific terms, Lovelock and an MIT colleague, Lynn Margulis, proposed that living organisms played a significant part in creating the earth's (and maybe Martian) atmosphere, and that these organisms act like the cells within a body to maintain a kind of homeostasis. Many environmentalists have decided that the earth *is* a living superorganism, even the Earth Mother.

Malthus, Thomas Robert. "An Essay on the Principle of Population as It Affects the Future Improvement of Society" [1798]. In *Voices of the Industrial Revolution,* John Bowditch and Clement Ramsland, eds. Ann Arbor: University of Michigan Press, Ann Arbor Paperback, 1961. In the early years of the Industrial Revolution, this Episcopal priest's essay distilled the fears that rapid increases in population would quickly consume the world's finite resources and overwhelm its ability to produce food. Malthus's vision remains the essential vision of the environmental movement.

Marx, Leo. *The Machine in the Garden: Technology and the Pastoral Ideal in America.* Oxford: Oxford University Press, 1964. No historian before or since has traced the themes that formed American attitudes toward nature. The "garden" refers to an ideal of orderly and friendly nature derived from preindustrial Europe. The "machine," of course, is industry. This 1964 book shows how we arrived at our present uneasy sense that industrial civilization and the natural world are incompatible.

Mitchell, John G., and Constance L. Stallings, eds. *Ecotactics: The Sierra Club Handbook for Environmental Activists.* New York: Pocket Books, 1970. With a foreword by Ralph Nader, *Ecotactics* was prepared for the first Earth Day as a catalog of social and political action. With the inclusion of the poet Gary Snyder, the club reached out to the alternative culture audience of the era.

Porter, Eliot, ed. *"In Wildness Is the Preservation of the World."* San Francisco: Sierra Club, 1962. With this book, the Sierra Club president David Brower launched the environmental movement's most successful publishing venture—the coffee-table environmental picture book. Porter's photographs taught millions of readers to look at nature up close, and he had the gift of making such things as caterpillars and their cocoons beautiful, rescuing them from the "pest" category imposed by utilitarian agriculture.

Rousseau, Jean Jacques. *The Social Contract* and *Discourse on the Origin of Inequality.* New York: Washington Square Press, 1973. These two essays are Rousseau's most influential works. The first is a critique of society at the end of the Enlightenment's great flowering of science and the beginning of the Industrial Revolution. The second is his theory of how to build a humane and egalitarian society.

Schumacher, Ernst F. *Small Is Beautiful: Economics as if People Mattered.* New York: Harper & Row, 1973. Many environmentalists know only the title of this book, which contains its guiding principle. Schumacher said, "I propose to give a new direction to technological development." That is what this book did for many government planners and world aid organizations. He proposed that small-scale "appropriate technologies" would do more for Third World economies than large-scale Western industry.

Simon, Julian, and Herman Kahn, eds. *The Resourceful Earth: A Response to "Global 2000."* New York: Basil Blackwell, 1984. This book did not get equal time with Barney's *Global 2000 Report to the President,* prepared for President Carter in 1980. *Global 2000* predicted "the world in 2000 will be more crowded, more polluted, less stable ecologically, and more vulnerable to disruption than the world we live in now. . . . Despite greater material output, the world's people will be poorer in many ways than they are today." Julian Simon of the University of Maryland and the late Herman Kahn of the Hoover Institution assembled distinguished experts to present a wealth of data to rebut this view, including many charts and graphs to show that there were few reasons to think we could not continue improving the conditions of human life. It is still the best documented survey of its kind.

Sitarz, Daniel. *Agenda 21: The Earth Summit Strategy to Save Our Planet.* Boulder, Colo.: Earthpress, 1993. The official summary of the 1992 United Nations Conference on Environment and Development has been edited for readability by the attorney Daniel Sitarz.

Thoreau, Henry David. *Walden.* New York: Grosset and Dunlap, 1910. Thoreau was a brilliant writer and a nineteenth-century middle-class

Romantic who, like many young Americans of the 1960s and 1970s, could not find a comfortable place in America. He still provides the rationale for alienated Americans who want nature to be an alternative to human society. A brilliant describer and a mediocre thinker.

White, Lynn. "The Historical Roots of Our Ecological Crisis." *Science* 155 (1967), 1203–7. The central idea of this essay is that the Judeo-Christian tradition is responsible for environmental destruction because it removed God from nature and gave humanity license to dominate the earth. Few environmentalists know the essay, but its central idea is widely paraphrased whenever someone writes about the connection between cultural values and environmental attitudes.

A BIBLIOGRAPHY OF WORKS CITED AND USED

Adler, Jonathan, H. *Little Green Lies: Environmental Miseducation.* Heritage Foundation, on-line electronic library.

Ahmad, Yusuf J., Salah El Serafy, and Ernst Lutz. *Environmental Accounting for Sustainable Development.* Washington, D.C.: World Bank, 1989.

Allen, Frank Edward. "School Children Learn Ecology by Doing—To Alleged Polluters." *Wall Street Journal,* April 30, 1992, A1.

Ames, Bruce. "Misconceptions about Pollution and Cancer." *National Review,* Dec. 3, 1990, 34–35.

———. "Natural Carcinogens and Dioxin." *SCI Total Environment* 104.(1–2) (May 1991): 159–66.

———. "Science and the Environment: Facts v. Phantoms." *National Wilderness Institute Resource* (spring 1993).

Anthony, Carl. "Eco-Justice." *Turning Wheel* (spring 1993): 18–22.

Archer, John S., ed. *Drawing the Line: Property Rights and Environmental Protection.* Olympia: Washington Research Council, 1992.

Arnold, Andrea. *Fear of Food: Environmentalist Scams, Media Mendacity and the Law of Disparagement.* Bellevue, Wash.: Free Enterprise Press, 1990.

Audubon (January 1993): 94–95.

Babbitt, Bruce. "The World after Rio." *World Monitor,* June 1992.

Baden, John A. "Tropical Tragedy." *National Review,* Oct. 1, 1990, 44.

Bailey, Ronald. "The Apocalypse Abusers." *CEI Update,* July 1993.

———. "Raining in Their Hearts." *National Review,* Dec. 3, 1990, 32–36.

Bandow, Doug, ed. *Protecting the Environment: A Free Market Strategy.* Washington, D.C.: Heritage Foundation, 1986.

Barnes, Fred. "What It Takes." *The New Republic,* Oct. 19, 1992, 22.

Barnet, Richard J. "No Room in the Lifeboats." *New York Times Magazine,* Apr. 16, 1978, 32–38.

Barney, Gerald O. "The Future of the Earth." *Union News* (winter 1990): 2.
————, ed. *The Global 2000 Report to the President: Entering the Twenty-first Century.* U.S. Government Printing Office, Washington, D.C.: CEQ and Department of State, 1980.
Barreiro, Jose. "Indigenous Peoples Are the 'Miner's Canary' of the Human Family." In *Learning to Listen to the Land*, edited by Bill Willers. Washington, D.C.: Island Press, 1991.
Becker, Carl. *The Heavenly City of the Eighteenth-Century Philosophers.* New Haven, Conn.: Yale Paperbound, 1960.
Beckwith, Jon. "A Historical View of Social Responsibility in Genetics." *Bio-Science* (May 1993): 327–33.
Berger, John J. *Environmental Restoration: Science and Strategies for Restoring the Earth.* Washington D.C.: Island Press, 1990.
Berry, Thomas. *The Dream of the Earth.* San Francisco: Sierra Club, 1988.
Berry, Wendell. *What Are People For?* San Francisco: Audio Literature, 1992.
————. "Word and Flesh." In *Helping Nature Heal*, edited by Richard Nilsen. Berkeley, Calif.: Whole Earth Catalog/Ten Speed Press, 1991.
"Beware of False Gods in Rio." *Wall Street Journal*, June 1, 1992.
Blessen, Richard P. "Wetlands Policy Still Unsettled: Interests Clash." *National Law Journal* 15 (Feb. 15, 1993), 25.
Blue-Print for Survival. The editors of *The Ecologist.* New York: Signet, 1974.
"Blood from a Turnip?" *Garden* 12: 26.
Bonynge, Francis. "The Future Wealth of America." New York, 1852. Reprinted in P. K. Whelpton, ed., "Population of the United States, 1925 to 1975." *American Journal of Sociology* 34, no. 2 (September 1928): 253–70.
Bookchin, Murray. "Death of a Small Planet." *The Progressive* 53 (August 1989): 19–23.
Borelli, Peter. "Whose Woods These Are I Think I Know—Robert Frost." *Amicus Journal* (winter 1981).
Bowditch, John, and Clement Ramsland, eds. *Voices of the Industrial Revolution.* Ann Arbor, Mich.: Ann Arbor Paperbacks, 1965.
Box, Thadis W. "Rangelands." In *Natural Resources for the 21st Century*, edited by R. Neil Sampson and Dwight Hair. Washington, D.C.: Island Press, 1990.
Brimelow, Peter, and Leslie Spencer. "You Can't Get There From Here." *Forbes,* July 6, 1992, 59–92.
Brookes, Warren T. "Dead Wrong Again." *National Review,* Oct. 7, 1991.
————. "The Strange Case of the Glancing Geese." *Forbes,* Sept. 2, 1991.
————. "War on Property Rights." *Washington Times,* op ed, July 18, 1991, G1.
Brooks, David. "Journalists and Others for Saving the Planet." *Wall Street Journal,* Oct. 5, 1989, A28.
Brooks, Paul. "Notes on the Conservation Revolution." In John G. Mitchell and Constance L. Stallings, eds. *Ecotactics: The Sierra Club Handbook for Environment Activists.* New York: Pocket Books, 1970.
Browder, John O. "The Limits of Extractivism: Tropical Forest Strategies

Beyond Extractive Reserves." *BioScience* (March 1992): 174–82.

Brown, Lester, Sandra Postel, and Christopher Flavin. "A World at Risk." In *State of the World 1989*, edited by Linda Starke. New York: W. W. Norton, 1970.

Browne, Malcolm W. "Cooling with Sound: An Effort to Save Ozone Shield." *New York Times*, Feb. 25, 1992, C1, C7.

Capano, E. "Eco-Activism and Junk Science." Town Hall electronic mail, Sept. 16, 1991, 0948 General.

Carson, Rachel. *Silent Spring*. Boston: Houghton Mifflin, 1962.

Catlin, George. *Letters and Notes on the Manners, Customs and Conditions of the North American Indians*. London: D. Bogue, 1842.

Chagnon, Napoleon A. *Yanomamo: The Fierce People*. New York: Holt, Rinehart and Winston, 1977.

———. "Yanomamo Survival." *Science*, April 7, 1989, 11.

Chakravaty, Subrata N. "Dean Buntrock's Green Machine." *Forbes*, Aug. 2, 1993.

Chase, Alston. "Property Rights Lose Out to Biocentrism." *PERC Reports* 11, no. 1 (March 1993).

Clawson, Marion. "Entering the Twenty-First Century—The Global 2000 Report to the President." *Resources* 66 (spring 1981).

Club of Rome. *The Limits to Growth: A Report for the Club of Rome's Project on the Predicament of Mankind*. New York: Universe Books, 1972.

Collard, André. *Rape of the Wild: Man's Violence Against Animals and the Earth*. Bloomington: Indiana University Press, 1990.

Commoner, Barry. *The Closing Circle: Nature, Man, and Technology*. New York: Knopf, 1971.

———. "Why We Have Failed." *Greenpeace* (September/October 1989).

Cowell, A. *The Decade of Destruction: The Crusade to Save the Amazon Rain Forest*. New York: Henry Holt, 1990.

Cronon, William. *Changes in the Land: Indians, Colonists and the Ecology of New England*. New York: Hill and Wang, 1984.

Crump, Andy. *Dictionary of Environment and Development: People, Places, Ideas and Organizations*. Cambridge, Mass.: MIT Press, 1993.

Cudmore, L. L. Larison. *The Center of Life*. New York: Quandrangle, 1977.

Cummings, Charles. *Eco-Spirituality*. Mahwah, N.J.: Paulist Press, 1991.

Daly, Herman E. "Free-Market Environmentalism: Turning a Good Servant into a Bad Master." *Critical Review* 6 nos. 2–3 (1993).

Daniel, Joseph E. "Revolution of the Right." *Buzzworm* 5 (May–June, 1993).

De Bell, Garrett, ed. *The Environmental Handbook*. New York: Ballantine Books, 1970.

Deloria, Vine, Jr. *God Is Red*. New York: Laurel Books, 1973.

Douglas, William O. *A Wilderness Bill of Rights*. Boston: Little, Brown, 1965.

Drummey, James J. "The Land Grabbers." *New American*, March 12, 1990.

Dukes, Carol Muske. "Evil Science Runs Amok—Again." *New York Times*, June 10, 1993, A27.

Dunn, James R. "America the Beautiful." *National Review*, July 6, 1992.

Echard, Jo Kwong. *Studies in Organization Trends: 5. Protecting the Environment: Old Rhetoric, New Imperatives.* Washington, D.C.: Capital Research Center, 1990.

Efron, Edith. *The Apocalyptics: Cancer and the Big Lie.* New York: Simon & Schuster, 1984.

Ehrenfeld, David. *Biological Conservation.* New York: Holt, Rinehart and Winston, 1970.

————. "Desert Life." *Orion* (spring 1991): 8.

Ehrlich, Paul. "Eco-Catastrophe!" *Ramparts* (September 1969).

————. *The Population Bomb.* New York: Ballantine Books, 1967.

Ehrlich, Paul, and Anne Ehrlich. *Healing the Planet.* Reading, Mass.: Addison Wesley, 1991.

Ehrlich, Paul R., and Edward O. Wilson. "Biodiversity Studies: Science and Policy." *Science* 253 (Aug. 16, 1991), 758–62.

Eisenberg, Evan. "Back to Eden." *Atlantic Monthly,* November 1989.

"Endangered Wildlife Species Face Extinction in China." *CanTibNet Newsletter,* Jan. 18, 1993, 26.

"EPA v. Private Property" (editorial). *Wall Street Journal,* Aug. 27, 1990.

Federal Manual for Identifying and Delineating Jurisdictional Wetlands (1989). Rev. 1991. Washington, D.C.: U.S. Government Printing Office, 1989.

Foster, Sarah. "Captain Planet Goes to Washington." *The Free Market* 11, no. 6 (June 1993).

"Founding Statement of the Audubon Society." *Audubon* 1, no. 1 (1886): 1.

Fox, Stephen. *The American Conservation Movement: John Muir and His Legacy.* Madison: University of Wisconsin Press, 1985.

Fremlin, J. H. *Power Production: What Are the Risks?* New York: Oxford University Press, 1986.

Fuller, R. Buckminster. *Utopia or Oblivion: The Prospects for Humanity.* New York: Bantam Books, 1969.

Gabriel, Trip. "Wildmen." *New York Times Magazine,* Oct. 14, 1990, 37–47.

Gardner, Robert, and Karl G. Heider. *Gardens of War: Life and Death in the New Guinea Stone Age.* New York: Random House, 1968.

George, Eric. *The Life and Death of Benjamin Robert Haydon, 1786–1846.* Oxford: Clarendon Press, 1967.

Goodman, Paul. "Self-Sufficiency Farming—1930's/1970's Style." In *The New Consciousness,* edited by Albert J. LaValley. Cambridge, Mass.: Winthrop, 1972.

Gomez-Pampa, Arturo, and Andrea Kaus. "Taming the Wilderness Myth." *BioScience* 42 (April 1992).

Gore, Al. *Earth in the Balance: Ecology and the Human Spirit.* Boston: Houghton Mifflin, 1992.

————. "An Ecological Kristallnacht. Listen." *New York Times,* Mar. 19, 1989, D27.

The Greenhouse Conspiracy. Video, available through the Conservative Video Club, 311 Massachusetts Ave., Washington, D.C. 20002.

Hair, Jay. "In Bear Country." *National Wildlife,* February 1988.

Hardin, Garrett. *Nature and Man's Fate.* New York: New American Library, 1961.

Harte, John, et al. *Toxics A to Z.* Berkeley: University of California Press, 1991.

Healy, Robert G. *Competition for Land in the American South: Agriculture, Human Settlement, and the Environment.* Washington D.C.: Conservation Foundation, 1985.

"Heidelberg Appeal to Heads of States and Governments," 3rd revision. Heidelberg, Germany, Apr. 14, 1992.

Herstgaard, Mark. "Covering the World: Ignoring the Earth." In *The Rolling Stone Environmental Reader.* Washington, D.C.: Island Press, 1992.

Hocking, Martin B. "Paper Versus Polystyrene: A Complex Choice." *Science* 251 (Feb. 1, 1991): 504–5.

Hoffman, Matthew. "Ratepayers Beware." *UpDate* (August 1993).

Holland, H. D. *The Chemical Evolution of the Atmosphere and the Oceans.* Princeton, N.J.: Princeton University Press, 1984.

Holloway, Mark. *Heavens on Earth: Utopian Communities in America, 1680–1880,* 2nd ed. New York: Dover, 1966.

Hudson, Wendy E., ed. *Landscape Linkages and Biodiversity.* Washington, D.C.: Island Press, 1990.

Huth, Hans, ed. *Nature and the Americans: Three Centuries of Changing Attitudes.* Lincoln: University of Nebraska Press, 1970.

"Instant Trees." *The Economist,* Apr. 28, 1990, 93.

Jackson, Norman W. "How Can You Buy or Sell the Sky?" *Connections* (October 1985): 11–12.

Javna, John. *Fifty Simple Things Kids Can Do to Save the Earth.* Berkeley, Calif.: Earth Works Group, 1989.

Jeffers, Susan. *Brother Eagle, Sister Sky: A Message from Chief Seattle.* New York: Dial Books, 1991.

Jevons, W. Stanley. *The Coal Question.* New York: Augustus M. Kelley, 1965.

Johnson, Allen. "How the Machiguenga Manage Resources: Conservation or Exploitation of Nature?" In A. Posey and W. Baloe, eds., *Advances in Economic Botany.* New York: New York Botanical Garden, 1989.

Johnson, Paul. *Intellectuals.* New York: HarperPerennial, 1990.

Johnson, Trebbe. "Caring for the Earth: New Activism Among Hopi Traditionals." *Amicus Journal* (winter 1991): 22.

Jones, Alex S. "Al Gore's Double Life." *New York Times Magazine,* Oct. 25, 1992, 40–79.

Jones, Lewis Thomas. *Aboriginal American Oratory.* Los Angeles: Southwest Museum, 1965.

Kaplan, David, and Bob Cohn. "Open Season on Environmental Laws." *Newsweek,* March 9, 1992.

Kiefer, Michael. "Fall of the Garden of Eden." *International Wildlife* 19, no. 4 (July/Aug. 1989): 39–43.

Koshland, Daniel E. "Scare of the Week." *Science* 244 (Apr. 7, 1989).

Krug, Edward C. "The Corrosion of Science." *Liberty* (March 1992): 29–30.

———. "Environmentalism: Abuse of a Just Cause." *Chronicles* (June 1993): 44–46.

Kulakov, Yuri, and Yuri Koropanchinskii. "The Path of Scientific Truth Leads to God." *Liberty* (June 1993): 26–31.

Laffer, William III. "Protecting Ecologically Valuable Wetlands Without Destroying Property Rights." Heritage Foundation (electronic bulletin board). July 15, 1991.

Lappe, Francis Moore, and Joseph Collins. *Food First.* London: Abacus, 1982.

Leal, Donald. "Unlocking the Logjam over Jobs and Endangered Animals." *San Diego Union-Tribune,* April 18, 1993.

Leal, Donald R., and Terry L. Anderson. "Buy That Fish a Drink." *Oregonian,* June 14, 1991.

Lehman, H. Jane. "Accord Ends Fight over Use of Land." *Washington Post,* July 17, 1993.

Libby, Lawrence W., and Rodney L. Clouser. "Population and Global Economic Patterns." In *Natural Resources for the 21st Century,* edited by R. Neil Sampson and Dwight Hair. Washington, D.C.: Island Press, 1990.

Locke, Timm. "Research Shows Consumers Concerned about Environmental Impacts of Purchasing Decision." News release, March 11, 1993, Western Wood Works.

Longman, K. A., and J. Jenik. *The Tropical Forest and Its Environment.* Harlow, England: Longman Scientific and Technical, 1987.

Lovelock, James. "The Earth as a Living Organism." In *Learning to Listen to the Land,* edited by Bill Willers. Washington, D.C.: Island Press, 1991.

Lutzenberger, Jose. "Environmental Ethics." *Surviving Together* (summer 1992): 8–10.

Lyons, Oren. "An Iroquois Perspective." In *American Indian Environments: Historical, Legal, Ethical, and Contemporary Perspectives.* Syracuse, N.Y.: Syracuse University Press, 1980.

McCoy, Bodie. "The Beginning or the End?" *The Networker* (February 1993): 7.

McHale, John. *World Facts and Trends.* New York: Macmillan, 1972.

McInnis, Doug. "Powder River Coal: Geologic Enigma, Environmental Dilemma." *Earth* (May 1993): 54–59.

McKibben, Bill. *The End of Nature.* New York: Random House, 1989.

Maddox, John. *The Doomsday Syndrome.* New York: McGraw-Hill, 1972.

Mahler, Halfdan. "People." *Scientific American* (Sept. 1980): 67–77.

"Man the Destroyer." *Audubon* 1, no. 1 (1886).

Mann, Charles C. "Extinction: Are Ecologists Crying Wolf?" *Science* 253 (1991): 736–38.

Margolis, Kenneth. "A Native Perspective." In the EcoTrust newsletter, 1992.

Marine, Gene. *America the Raped: The Engineering Mentality and the Devastation of a Continent.* New York: Simon & Schuster, 1969.

Markham, James M. "Paris Groups Urge 'Decisive Action' for the Environment." *New York Times,* July 17, 1989.

Marsh, George Perkins. *Man and Nature, or Physical Geography as Modified by Human Action*. Cambridge, Mass.: Harvard University Press, 1965.

Marshall, Robert. *The People's Forests*. New York: Harrison Smith and Robert Haas, 1933.

Martin, Laura C. *The Folklore of Trees and Shrubs*. Chester, Conn.: Globe Pequot Press, 1992.

Marx, Karl, and Friedrich Engels. *Communist Manifesto*. Hammondsworth: Penguin, 1967.

Marx, Leo. *The Machine in the Garden: Technology and the Pastoral Ideal in America*. Oxford: Oxford University Press, 1964.

Maser, Chris. *Forest Primeval*. San Francisco: Sierra, 1989.

Milton, Katharine. "Civilization and Its Discontents: Amazon Indians Experience the Thin Wedge of Materialism." *Natural History* (March 1992): 37–42.

Miniter, Richard. "Out to Change the Law of the Land." *Insight*, May 17, 1993.

Mitchell, John G., and Constance L. Stallings, eds. *Ecotactics: The Sierra Club Handbook for Environment Activists*. New York: Pocket Books, 1970.

Molina, Anatilde Idoyaga. "Pilaga Seduction Songs." *Latin American Indian Literatures Journal* 5, no. 2 (fall 1989).

Monastersky, Richard. "The Deforestation Debate." *Science News*, July 10, 1993, 26–27.

Monks, Vicki. "See No Evil." *American Journalism Review* (June 1993): 18–25.

Moon, William Least Heat. *Blue Highways*. Boston: Little, Brown, 1982.

Morrison, Dan. "In the Belly of the Crocodile." *American Way* (Feb. 1, 1992): 53–93.

Nash, Hugh. *Progress As If Survival Mattered: A Handbook for a Conserver Society*. San Francisco: Friends of the Earth, 1977.

Nations, James D., and Ronald B. Nigh. "The Evolutionary Potential of Lacandon Maya Sustained-Yield Tropical Forest Agriculture." *Journal of Anthropological Research* (spring 1980): 1–28.

Natural Resources Defense Council. *Fifty Simple Things You Can Do to Save the Earth*. Berkeley, Calif.: Earth Works Group, 1989.

Neiring, William. *Wetlands of North America*. Charlottesville, Va.: Thomasson-Grant, 1991.

Nelson, Ingrid. "Green Hills, Blue Lakes, and Red Tape." *Country Journal* (September 1986).

"The New Explorers." Public Television, Feb. 5, 1991.

Nietzsche, Friedrich. *Beyond Good and Evil: Prelude to a Philosophy of the Future*. New York: Russell and Russell, 1964.

Nilsen, Richard, ed. *Helping Nature Heal*. Berkeley, Calif.: Ten Speed Press, 1992.

Noyes, Russell, ed. *English Romantic Poetry and Prose*. New York: Oxford University Press, 1956.

Olcott, Martha Brill. *The Kazakhs*. Stanford, Calif.: Hoover Institution Press, 1987.

Panczenko, Oleg. "A Ridiculous Degreed Ass." *Town Hall* (electronic bulletin board). April 22, 1992.

Parker, Tony, and Robert Allerton. *The Courage of His Convictions.* London: Hutchinson, 1962.

Patton, Benjamin W. "Who's Endangered?" *NWI Resource* 2, 3 (fall 1991).

Pawlick, Thomas. *A Killing Rain: The Global Threat of Acid Precipitation.* San Francisco: Sierra Club, 1984.

Peters, Gary L., and Robert P. Larkin. *Population Geography: Problems, Concepts and Prospects.* Dubuque, Iowa: Kendall Hunt, 1983.

Peterson, D. J. *Troubled Lands: The Legacy of Soviet Environmental Destruction.* Boulder, Colo.: Westview Press, 1993.

Peterson, Nicholas, ed. *Aboriginal Land Rights.* Canberra: Australian Institute of Aboriginal Studies, 1981.

Picard, André. "James Bay II." *The Amicus Journal* (fall 1990).

Pinkalla, Diane. "Were Pygmy Hippos Hunted to Extinction?" *Earth* (May 1993): 16–17.

Plotkin, Mark J., and Lisa M. Famolare, eds. *Sustainable Harvest and Marketing of Rain Forest Products.* Washington, D.C.: Island Press, 1992.

Plucknett, Donald L. "International Agricultural Research for the Next Century." *BioScience* (July/August 1993): 432–40.

Ponting, Clive. *A Green History of the World.* London: Sinclair-Stevenson, 1991.

Posey, A., and W. Baloe, eds. "Resource Management in Amazonia: Indigenous and Folk Strategies." In *Advances in Economic Botany,* vol. 7. Bronx, N.Y.: Bronx Botanical Garden.

Redford, Kent H. "The Ecologically Noble Savage." *CS Quarterly* 15 (1991): 46–48.

Reich, Charles A. *The Greening of America.* New York: Random House, 1970.

Reiger, John F. *American Sportsmen and the Origins of Conservation,* rev. ed. Norman: University of Oklahoma Press, 1986.

Reigle, Margaret Ann. "Environmentalists Concerned by Growing Grassroots Backlash." *News from the FLOC* (April 1993).

"Review and Outlook." *Wall Street Journal,* August 27, 1990, A10.

Ridley, Matt, and Bobbi S. Low. "Can Selfishness Save the Environment?" *Atlantic Monthly* 272, no. 3 (September 1993): 76–86.

Rolling Stone Environmental Reader. Washington, D.C.: Island Press, 1992.

Rosenberg, Nathan, and L. E. Birdzell. *The Economic Transformation of the Industrial World.* New York: Basic Books, 1986.

Salomon, Michel. "Heidelberg to Rio: Itinerary of an Approach." *Projections* (1992).

Sampson, R. Neil, and Dwight Hair, eds. *Natural Resources for the 21st Century.* Washington, D.C.: Island Press, 1990.

Sayres, Sohnya. "Rape of the Wild." *Amicus Journal* (fall 1990): 39–42.

Schlesinger, M. E., ed. *Greenhouse-Gas-Induced Climatic Change: A Critical Appraisal of Simulations and Observations.* Amsterdam: Elsevier, 1991.

Schneider, Keith. "Environmental Laws Face a Stiff Test from Landowners," *New York Times.* Jan. 20, 1992, A1.

Schroeder, Herb. "Symbolism and Spiritual Values in Experiencing Nature." North Central Forest Experiment Station Seminar, Jan. 15, 1992.

The Science of Wetland Definition and Delineation. Hearing Before the Subcommittee on Environment of the Committee on Science, Space and Technology, 102nd Congress, first session, November 12, 1991. Washington, D.C.: U.S. Government Printing Office, 1992.

Sheehy, Gail. "Gore—The Sun Also Rises." *Vanity Fair* (March 1988): 141–98.

Shiva, Vandana. *The Violence of the Green Revolution.* London: Zed Books, 1991.

Shoumatoff, Alex. *The World Is Burning.* Boston: Little, Brown, 1990.

Sidey, Hugh. "An Opposing View." *Time* 127 (Feb. 24, 1986): 21.

Simmons, Randy, and Urs Kreuter. "Save an Elephant—Buy Ivory." *Washington Post,* Oct. 1, 1989, D3.

Sitarz, Daniel, ed. *Agenda 21: The Earth Summit Strategy to Save Our Planet.* Boulder, Colo.: Earth Press, 1993.

Smith, Fred L. "Carnival of Dunces." *National Review,* July 6, 1992.

Smith, Harry Lee. "The Environment Since the Industrial Revolution." *Cato Policy Report* 13 (March/April 1991).

Smithsonian Institution. *The Global Environment: Are We Overreacting?* Washington, D.C.: 1989.

Snow, Ed. *Inside the Environmental Movement.* Washington, D.C.: Island Press, 1992.

Snyder, Gary. "The Etiquette of Freedom." *Sierra* (September/October, 1989): 114.

Spengler, Oswald. *The Decline of the West,* ed. and trans. Charles Francis. New York: Knopf, 1926.

———. *Der Untergang des Abendlandes: Umrisse einer Morphologie der Weltgeschichte.* München: D. H. Beck, 1973.

Stevens, William K. "Balance of Nature? What Balance is That?" *New York Times,* Oct. 22, 1991, C4.

Stone, R. "Babbitt Shakes Up Science at Interior." *Science* 276 (1993): 976–78.

Stroup, Richard L. "Global Warming: U.S. Should Stay Cool." *Seattle Times,* May 24, 1992.

———. "Reform Endangered Species Act: Pay Landowners Rather Than Penalize Them." PERC News Release, May 17, 1993.

———. "Voters Need to Understand the Importance of Property Rights." *PERC Reports* 11 (March 1993).

Stuart, George E. "Maya Heartland Under Seige." *National Geographic* (November 1992): 94.

Stutz, Bruce. "The Landscape of Hunger." *Audubon* (March/April 1993): 54–63.

"Summit of the Arch." *Department of State Bulletin* 89, no. 2150 (September 1989): 1.

Talbot, Lee M. "Man's Role in Managing the Global Environment." In *Changing the Global Environment,* edited by Daniel B. Botkin et al. (Boston:

Academic Press, 1989).

Thomas, Lewis. "Crickets, Cats, Bats, and Chaos." *Audubon* (March/April 1992): 94–99.

———. *The Lives of a Cell: Notes of a Biology Watcher.* New York: Viking, 1974.

Thoreau, Henry David. *The Portable Thoreau,* ed. Carl Bode. New York: Viking, 1947.

Tilman, David, and David Wedin. "Oscillation and Chaos in the Dynamics of a Perennial Grass." *Nature* 353 (Oct. 17, 1991): 653–55.

Tolan, Sandy, and Nancy Postero. "Accidents of History." *New York Times Magazine,* Feb. 23, 1992, 38.

Trexler, Mark C. *Minding the Carbon Store: Weighing U.S. Forest Strategies to Slow Global Warming.* Washington, D.C.: World Resources Institute, 1991.

Vecsey, Christopher, and Robert W. Venables, eds. *American Indian Environments.* Syracuse, N.Y.: Syracuse University Press, 1980.

Wald, George. "There Isn't Much Time." *The Progressive* (December 1975): 22–24.

Weilbacher, Mike. "Earth Day: The New Children's Crusade." *E Magazine* (March/April 1993): 31–35.

Wexler, Mark. "Sacred Rights." *National Wildlife* (June 1992): 18–23.

Whelpton, P. K. "Population of the United States, 1925 to 1975." *American Journal of Sociology* 34, no. 2 (September 1928): 253–270.

White, Lynn, Jr. *Medieval Technology and Social Change.* New York: Oxford University Press, 1966.

Wild, Peter. *Pioneer Conservationists of Eastern America.* Missoula, Mont.: Mountain Press, 1986.

Willers, Bill, ed. *Learning to Listen to the Land.* Washington, D.C.: Island Press, 1991.

Wilson, E. O., ed. *Biodiversity.* Washington, D.C.: National Academy Press, 1988.

———. *Biophilia: The Human Bond with Other Species.* Cambridge, Mass.: Harvard University Press, 1984.

Witt, Shirley Hill, and Stan Steiner. *The Way: An Anthology of American Indian Literature.* New York: Knopf, 1972.

Wollstonecraft, Mary. *The Vindication of the Rights of Woman.* London: J. M. Dent and Sons, [1792] 1929.

World Resources Institute. *World Resources, 1992–1993.* New York: Oxford University Press, 1992.

"Worried Memo: Dems Ponder Green Gore." *Wall Street Journal,* Sept. 13, 1992, A14.

Wright, Sigrid. "'Property Rights' Issue Re-Ignited." *The Leader,* April 1993.

Yoffe, Emily. "Silence of the Frogs." *New York Times Magazine,* Dec. 13, 1992.

Zwick, David, and Marcy Benstock. *Water Wasteland.* New York: Grossman, 1971.

INDEX